BARRON'S
E-Z
BUSINESS
MATHEMATICS

Calman Goozner
Thomas P. Walsh, Ed.D., Kean University

BARRON'S

650.0151
Goo

All inquiries should be addressed to:
Barron's Educational Series, Inc.
250 Wireless Boulevard
Hauppauge, New York 11788
www.barronseduc.com

Library of Congress Catalog Card No. 2009008720
ISBN-13: 978-0-7641-4259-8
ISBN-10: 0-7641-4259-3

Library of Congress Cataloging-in-Publication Data
Goozner, Calman.
  E-Z business mathematics / Calman Goozner and Thomas P. Walsh.—4th ed.
    p.   cm.—(E-Z)
  Includes index.
  "Previous editions under the title: Business mathematics the easy way."
  ISBN-13: 978-0-7641-4259-8
  ISBN-10: 0-7641-4259-3
  1. Business mathematics.   I. Walsh, Thomas P.   II. Goozner, Calman. Business mathematics
the easy way.   III. Title.   IV. Title: Easy business mathematics.
  HF5691.G655 2009
  650.01'513—dc22

                                        2009008720

Printed in the United States of America
9 8 7 6 5 4 3 2 1

ACKNOWLEDGMENTS

I would like to thank my father and mother, Robert and Mary, for their support and love throughout the years in my educational pursuits. They encouraged all of their children to get as much education as possible, and it was the best start their children (Dan, Pat, Tom, Bob, Steve, Jean, and John) could have gotten.

To the student: Read before you begin this book

Math is not the easiest of subjects to master, but you already know that. You've probably bought this book to get good (or, at least decent) in the math needed for business. The book is written in a way that encourages you to go to any chapter and jump in; you don't need to get through earlier chapters before you attempt later chapters. Keep one thing in mind as you go through these lessons, however: Math is how we make sense of the world around us and order it, to a great extent. Math is the study of patterns in our world, and once we find the pattern, we write it as an equation (or formula). Understanding these formulas, and these concepts, will help you to predict what will happen in the future, and thereby profit from your predictions. You are a better consumer, and a better businessperson, if you can use these concepts to buy more with less, and to save more of your money.

CONTENTS

Fundamental Arithmetic and Algebra Review

Unit 1: Addition in Business

In business, you will use addition every day to be effective. Whether you are calculating how many items in your store you have sold, working up a sales slip or an invoice, calculating the total of a bill (with tip) in a restaurant, or checking a time card, addition skills are critical to your success.

REVIEW OF ADDITION

Fortunately, addition is the easiest of operations to do. Addition is when two or more numbers (called the *addends*) are added together. The result is called the *sum*.

EXAMPLE 1

$$\begin{array}{r} \overset{1\ 1}{563\}} \quad \Leftarrow \textbf{addends} \\ +\ 874\} \quad \Leftarrow \textbf{addends} \\ \hline 1437 \quad \Leftarrow \textbf{sum} \end{array}$$

You probably will not see the word *addend* ever in business, but you will see the word *sum*. Always line up the columns: the ones, the tens, the hundreds, and so forth.

EXAMPLE 2

$$\begin{array}{r} \overset{1\ 1\ 1}{982} \\ 415 \\ +\ 377 \\ \hline 1774 \end{array}$$

Most of the time, you will certainly use a calculator to obtain these sums (Indeed, very few business people do math by hand.) However, knowing how addition is done may be helpful. After all, you might find yourself without a calculator one day.

EXAMPLE 3

$$
\begin{array}{r}
{}^{2}{}^{2}82 \\
53 \\
44 \\
18 \\
+37 \\
\hline
234
\end{array}
$$

Consider Example 3, above. Many accountants (who need to add large columns of numbers) use this little shortcut: they make groups of 10. Look at the ones column. Pair up 8 and 2 to make 10, pair up 7 and 3 to make another 10, and you are left with a 4. The 4 comes down, and you have 2 to carry. In the tens column, you can pair up 2 and 8, and then add 5, 4, and 1 to make two groups of 10. You are left with 3, which comes down. The 2 is carried to the hundreds place and brought down.

When adding money, line up the decimal places. This is how all decimal addition is carried out: line up the decimal places.

EXAMPLE 4

$$
\begin{array}{r}
{}^{1}\ \ {}^{1}\$673.85 \\
+\ \$51.51 \\
\hline
\$725.36
\end{array}
$$

CHECKING ADDITION

You should always check your addition, and it is easy to do with a calculator. The easiest way to check your addition is to add up the numbers in *reverse order*. That way, if the two sums you get agree, the chances are that you got it right. If the two sums do not agree, then you might consider adding the numbers again to find out which is the correct sum.

CALCULATOR OPTION

Checking an addition problem with a calculator is easy. In fact, not too many people do calculations on paper. Most use calculators. To check example 4 above, simply type in the numbers:

The individual keystrokes for the numbers and symbols are boxed. Some calculators will show dollar signs in the display, but most will not.

SPREADSHEET OPTION

A computer gives you another option for addition: a computer-based spreadsheet. Putting numbers into a computer spreadsheet can be easy, but it also can be time consuming. You must judge whether the time setting up the computer spreadsheet is worth the benefits in calculating the sum that way.

EXAMPLE 3 AGAIN

Here is example 3 on a spreadsheet.

82
53
44
18
37
234

The numbers were entered into the spreadsheet. The SUM function was used. The function looks like this on the standard toolbar: Σ.

When you use a spreadsheet, you must be *extra careful* to make sure the numbers have been entered correctly. If not, you might get an erroneous result and lose the business deal (worth millions, perhaps). It has happened in business many times before. **Always double-check your work**.

EXERCISES

Try these exercises on a calculator, on paper, *and* on a spreadsheet to determine which method you prefer.

Exercise A Solve the following problems, indicating zero cent where necessary, and check your answers:

1. $63.75 + $28.15 + $35.82 + $73.47

2. $232.43 + $364 + $472.10 + $527

3. $607.75 + $28 + $327.85 + $215

4. $48.73 + $2,070 + $628.47 + $372

5. $35 + $347.80 + $8,923

6. $4,623.49 + $793.78 + $93 + $78.50

7. $649.57 + $84 + $5,627.26 + $10,738.47

8. $6,327.56 + $14,246.76 + $976 + $18,648.95

9. $6,456.36 + $49,275.49 + $876 + $9,786.68

10. $5,458.63 + $7,568 + $43,786.17 + $596.56

Exercise B   Find the total for each of the following forms, and check your answers:
Sample Problem

11.

Expense Records Week of 9/19	
Salesperson	Amount
Anderson, B.	$63 | 50
Bates, G.	71 | 68
Berger, M.	94 | 32
Carter, J.	87 | 55
Chambers, V.	74 | 19
Total	$391 | 24

Check
63⬚5
⊞
71⬚68
⊞
94⬚32
⊞
87⬚55
⊞
74⬚19
⊟
✓391.24

12.

Expense Records Week of 10/3	
Salesperson	Amount
Anderson, B.	$88 | 50
Bates, G.	91 | 50
Berger, M.	84 | 82
Carter, J.	77 | 44
Chambers, V.	64 | 29
Total	

13.

	Date: September 3			
Salesperson	Sales		Commission	
Anderson, B.	$15,685	75	$1,568	58
Bates, G.	13,856	80	1,385	68
Berger, M.	18,560	44	1,856	04
Carter, J.	14,375	86	1,437	59
Chambers, V.	15,963	45	1,596	35
Totals				

14.

Inventory	
Item	Quantity
25-Watt Bulbs	75
40-Watt Bulbs	123
60-Watt Bulbs	115
75-Watt Bulbs	98
100-Watt Bulbs	135
Total	

15.

Inventory	
Item	Quantity
Men's Ties #62	2,478
Men's Ties #68	987
Men's Ties #71	3,628
Men's Ties #74	5,870
Men's Ties #78	697
Men's Ties #81	874
Men's Ties #85	1,320
Men's Ties #90	2,054
Total	

Find the subtotal and the total for the following form:

16.

HANDY & SONS

March 28 20--

SOLD TO _P. Griffiths_

ADDRESS _1031 Brookfield Ave._

CLERK _M.W._   DEPT _33_   AMT REC'D $_105_

QUAN.	DESCRIPTION	AMOUNT	
3	Sport Shirts	$26	85
6	Pair of Socks	11	70
2	Ties	18	00
1	Pair of Shoes	36	95
	Subtotal		
	8% Sales tax	7	48
	Total		

POSITIVELY NO EXCHANGES MADE UNLESS
THIS SLIP IS PRESENTED WITHIN 3 DAYS.

Word Problems

Word problems are easier to solve if you watch for key phrases. For instance, key phrases indicating addition are:

"Find the total."

"What is the cost of . . .?"

Use addition if you are given a set of numbers in the same units (money, hours, parts, etc.) and are asked for the *total,* the *sum,* or the *(total) cost.*

Exercise C Solve the following problems:

17. A company bought two typewriters for $1,525 each, a desk for $785.90, a chair for $124.75, and an adding machine for $1,250.95. Find the total amount of the purchases.

18. The Taylor Building Company deposited the following checks: $315.80, $475.60, $115.28, $287.60, $330.50, and $98.15. What was the total amount of the deposit?

19. Sam Steinfeld has the following deductions from his weekly paycheck: federal withholding tax, $42.50, Social Security, $12.75; state tax, $14.10; city tax, $8.75; medical insurance, $2.40. Find the total of his payroll deductions.

20. The breakdown of the Alvarez family's monthly mortgage payment is as follows: reduction of principal, $98.75; interest charges, $138.63; escrow for real estate tax, $113.78; and mortgage insurance premium, $13.52. How much is the total monthly payment?

21. Sharon earns part of her salary from commissions based on sales. Last week she earned the following commissions: $35.63, $41.50, $23.78, $47.83, and $38.35. She also earns a base salary of $125, to which the commissions are added. What was Sharon's total salary last week?

HORIZONTAL AND VERTICAL ADDITION

Many business forms are designed with sums to be added *horizontally* (across the page) and totaled in the right-hand column. Very often, these forms also require *vertical* addition, totaled under each column. This is an automatic check on your accuracy. If the two totals do not agree, you have made a mistake in addition.

CHECKING HORIZONTAL ADDITION

As in vertical addition, you should check your accuracy by adding the numbers in the reverse direction.

EXAMPLE Find the weekly sales by department and the grand total for the week of 1/15.

WEEKLY SALES BY DEPARTMENT

Dept.	Mon.	Tues.	Wed.	Thurs.	Fri.	Sat.	Sun.	Total
Grocery	$1,821 72	$1,763 15	$1,948 65	$2,163 47	$2,065 38	$1,918 42	$1,621 40	$13,302 19
Produce	963 54	907 43	1,163 70	1,485 72	1,391 72	1,268 34	1,020 65	8,201 10
Dairy	1,461 70	1,368 50	1,485 61	1,538 45	1,465 68	1,371 52	1,169 37	9,869 83
Meat	1,235 85	1,173 85	1,345 57	1,478 45	1,315 10	1,219 05	1,129 15	8,897 02
Deli	863 50	743 78	968 43	1,019 25	982 15	872 38	784 19	6,233 68
Nonfood	734 68	715 93	853 19	916 42	868 58	753 62	642 35	5,484 77
Totals	$7,080 99	6,672 64	7,765 15	8,601 76	8,088 61	7,403 33	6,367 11	51,979. 59 **Grand Total**

STEP 1. Using a calculator
    a. Add the daily sales and write each total in the total column on the right.
    b. Add each column and write the total under the column being added.
NOTE: c. Enter all decimal points.
STEP 2. As a check on accuracy, the two totals should be the same.
CHECK: $51,979.59 = $51,979.59 ✓

EXERCISES

Exercise D   Using a calculator, find the sum of each set of numbers by adding horizontally.

22. $8 + 5 + 3 + 2 + 8 + 9 =$

23. $32 + 45 + 13 + 72 + 63 =$

24. $53 + 61 + 94 + 68 + 47 =$

25. $3,427 + 5,628 + 7,964 + 4,943 =$

26. $23,472 + 47,065 + 17,528 =$

27. $94,128 + 21,072 + 15,735 =$

28. $96,725 + 63,420 + 15,823 =$

29. $50,623 + 18,972 + 21,682 =$

30. $23,125 + 48,575 + 63,945 + 15,075 =$

Exercise E   Find the totals on the following forms, and check your answers:

31.

Employee	\multicolumn{6}{c}{Hours Worked Week of 10/17}
	Mon.	Tues.	Wed.	Thurs.	Fri.	Total Hours
Adams, C.	8	8	12	12	10	_____
Adman, M.	9	9	13	4	10	_____
Burke, W.	8	8	10	8	7	_____
Curtis, A.	8	6	12	8	10	_____
Dellman, A.	12	8	6	12	6	_____
Evans, P.	8	10	8	8	12	_____
Totals	+	+	+	+	=	Grand Total

32.

\multicolumn{8}{c}{Payroll Deductions   Week of 3/23}

Employee Card No.	Federal Tax	FICA Tax	State Tax	City Tax	Pension	Health Plan	Total
01	$35 15	$12 25	$ 8 79	$ 5 18	$7 35	$2 15	_____ __
02	43 70	13 47	9 16	5 78	8 38	3 21	_____ __
03	42 61	14 18	9 83	6 10	8 74	3 47	_____ __
04	38 61	13 17	8 91	5 83	7 84	2 65	_____ __
05	53 26	19 25	12 42	7 86	9 24	3 81	_____ __
06	51 80	18 72	12 39	7 47	9 18	3 34	_____ __
07	45 73	16 37	9 72	6 19	7 49	2 73	_____ __
08	57 20	19 38	13 24	8 86	9 42	3 68	_____ __
09	38 90	13 41	8 64	5 93	7 92	2 71	_____ __
10	55 30	18 15	13 05	8 74	8 19	3 42	_____ __
Totals		+	+	+	+	+	=	Grand Total

33.

Dept.	Mon.		Tues.		Wed.		Thurs.		Fri.		Sat.		Sun.		Total	
Grocery	$1,943	16	$1,948	65	$1,863	41	$2,158	17	$1,984	53	$1,928	31	$1,730	09	_____	___
Produce	983	75	973	18	867	38	968	38	828	16	793	82	834	15	_____	___
Dairy	1,575	68	1,562	84	1,471	35	1,420	93	1,420	93	1,341	72	1,286	94	_____	___
Meat	1,316	37	1,347	19	1,285	64	1,241	63	1,241	63	1,163	58	1,092	66	_____	___
Deli	965	53	860	93	763	28	738	40	738	40	629	40	643	19	_____	___
Nonfood	847	19	784	52	654	32	635	19	635	19	515	15	553	55	_____	___
Totals		+		+		+		+		+		+		=		

Weekly Sales by Department Date: 1/8

Grand Total

34.

Department Sales Week of 6/17

Department	Cash		Charge		C.O.D.		Total	
Ladies' Wear	$3,425	65	$2,315	78	$3,478	58	_____	___
Men's Wear	2,975	40	1,848	37	2,264	37	_____	___
Children's Wear	1,563	90	987	35	1,287	50	_____	___
Appliances	2,562	76	4,628	74	3,768	87	_____	___
Furniture	3,728	35	5,947	39	4,215	95	_____	___
Toys	1,584	72	867	48	1,367	58	_____	___
Totals		+		+		=		

Grand Total

Exercise F Find totals on the following forms:

35. Please endorse all checks and list below singly

DATE	, 20	DOLLARS	CENTS
	BILLS	394	00
	COINS	19	59
	CHECKS	296	42
		358	47
		963	98
		152	63
	TOTAL		

36. Please endorse all checks and list below singly

DATE	, 20	DOLLARS	CENTS
	BILLS	474	00
	COINS	48	54
	CHECKS	939	75
		362	59
		691	35
		543	91
		478	79
		369	75
		592	34
	TOTAL		

37.
Please endorse all checks and list below singly

DATE	, 20	DOLLARS	CENTS
BILLS		893	00
COINS		5	98
CHECKS		692	65
		473	95
		230	77
		149	02
TOTAL			

38.
Please endorse all checks and list below singly

DATE	, 20	DOLLARS	CENTS
BILLS		190	00
COINS		4	55
CHECKS		39	98
		99	59
		46	22
		70	95
		68	87
		27	77
TOTAL			

39.

Petty Cash Expenditures

Date	Amount	
1/3	$22	50
1/4	39	68
1/5	34	99
1/6	8	77
1/7	28	66
1/10	43	42
1/11	3	19
1/12	17	50
1/13	21	63
Total		

40.

Petty Cash Expenditures

Date	Amount	
3/13	$28	50
3/14	19	68
3/15	34	79
3/16	8	18
3/17	43	65
3/19	22	42
3/20	3	17
3/21	17	50
3/22	21	63
Total		

41.

Petty Cash Expenditures

Date	Amount	
5/1	$20	65
5/2	31	19
5/3	5	78
5/4	14	60
5/5	37	92
5/7	28	43
5/8	12	21
5/9	19	46
5/10	25	28
Total		

Unit 2: Subtraction in Business

As with addition, subtraction skills are used every day in business.

REVIEW OF SUBTRACTION

Subtraction is quite easy. In fact, it is the inverse of addition. Usually it takes place when a smaller number is taken away from another, larger number. The larger number is called the *minuend*, and the smaller number is called the *subtrahend*. The result is called the *difference*.

$$
\begin{array}{rl}
683\} & \Leftarrow \textbf{minuend} \\
-251\} & \Leftarrow \textbf{subtrahend} \\
\hline
432 & \Leftarrow \textbf{difference}
\end{array}
$$

You probably will not ever see the words *minuend* and *subtrahend*, but you may see the word *difference* in business. In Example 1, no borrowing was involved. That is, no numbers in the subtrahend were bigger than the numbers in the minuend, so each individual number was subtracted (ones, tens, and hundreds). In this next example, there will be borrowing:

$$
\begin{array}{rl}
{}^{7\ 15\ 14} \\
3,864\} & \Leftarrow \textbf{minuend} \\
-597\} & \Leftarrow \textbf{subtrahend} \\
\hline
3,267 & \Leftarrow \textbf{difference}
\end{array}
$$

In this example, we tried to subtract 7 from 4 in the ones column, but we cannot do that. So we had to borrow 10 ones from the tens place and change 6 tens to 5 tens. Then we had 14 ones. We can subtract 7 from 14 to get 7. Moving over to the tens place, we try to subtract 9 from 5, but again we cannot do that. So we borrow 10 tens from the hundreds place, making the 8 as 7. Then we have 15 tens. Subtracting 9 from 15 gives 6. Moving to the hundreds place, we subtract 5 from 7 and get 2. Finally, we did nothing in the thousands place, so that the 3 comes down to the answer. The answer is 3267. Note we subtract only one place at a time. This makes the calculation fairly simple.

The next example uses money. This will show how decimals are handled. We subtract $45.60 from $72.35.

$$
\begin{array}{rl}
{}^{6\,11\ 13} \\
\$72.35 \\
-\$45.60 \\
\hline
\$26.75
\end{array}
$$

Again, as in addition, the decimal points are lined up. You must line them up every time. You will not get the right answer otherwise. Start the subtraction with the hundredths (right-most numbers). We subtract 0 from 5 and get 5. Next, subtract the tenths (next column to the left). We cannot take 6 from 3, so we borrow 1 from the 2 in the one place and then 6 from 13 is 7. Now, in the ones place, we cannot take 5 from 1, so again we borrow 1 from the tens place and make the 7 a 6. Then 5 from 11 is 6, and that takes care of the ones place. Finally, in the tens place, we take 4 from 6 and get 2. So the answer is $26.75.

For the next example, we will subtract two large-dollar amounts: $12,637,495.80 minus $5,265,296.65

EXAMPLE 4

$$\begin{array}{r} {}^{0\ 12\ 5\ 13\quad 31815\ 710} \\ \$12,637,495.80 \\ -\$\ \ 5,265,296.65 \\ \hline \$\ \ 7,372,199.15 \end{array}$$

The individual subtractions are clear. Only one individual subtraction is of note here—in the tens place. Subtracting 9 from 9 would typically give a 0. However, since we needed to borrow from the 9 (to give 15 to the ones column), the tens column now contained an 8. We then needed to borrow from the 4 in the hundreds column.

For a last example, we visit a condition that happens quite often in business: a deficit. What happens when you have some money in your bank account and you need to subtract *more* than what you have? In that case, you will end up with a deficit. For example, if you have $48,135.45 in a bank account but a bill for $50,000.00 comes due, what happens? What happens is that the bank will subtract the amount you have from the amount of the bill. The result becomes a deficit you will need to pay back to the bank.

EXAMPLE 5

$$\begin{array}{r} {}^{4\ 9\ 9\ 9\ 9\ 910} \\ \$50,000.00 \\ -\$48,135.45 \\ \hline \$1,864.55 \end{array}$$

Note that when you have a round amount ($50,000.00), you will change all the individual numbers to a number you can subtract from and then make the individual subtractions. So in this example, you would owe the bank $1,864.55. Most banks call this a deficit in your account and show it on your bank statement with a negative sign in front. Sometimes the deficit is shown with red ink, hence the term "in the red" for a deficit. Most banks will charge you a fee for having the deficit.

CHECKING SUBTRACTION

As noted with addition, you should always check your subtraction. It is easy to do with a calculator. The easiest way to check subtraction is to add the difference and the subtrahend up to see if the result equals the minuend. If it does, you probably have the right answer. If it does not equal the original minuend, then you might consider doing the subtraction again to find out which is the correct answer.

EXAMPLE 6

$$
\begin{array}{r}
\substack{317\ 611} \\
\$347.71 \\
-\$239.55 \\
\hline
\$108.16
\end{array}
$$

Example 6 shows the subtraction. Now check by adding the difference and the subtrahend:

$$\$108.16 + \$239.55 = \$347.71$$

The result equals the minuend, so it checks out.

EXERCISES

Exercise A Find the difference in each of the following problems, and check your answers:

1. $\begin{array}{r}65\\-23\\\hline\end{array}$	2. $\begin{array}{r}96\\-53\\\hline\end{array}$	3. $\begin{array}{r}76\\-56\\\hline\end{array}$	4. $\begin{array}{r}98\\-49\\\hline\end{array}$	5. $\begin{array}{r}825\\-768\\\hline\end{array}$

6. $\begin{array}{r}328\\-249\\\hline\end{array}$ 7. $\begin{array}{r}738\\-649\\\hline\end{array}$ 8. $\begin{array}{r}\$537.54\\-309.19\\\hline\end{array}$ 9. $\begin{array}{r}\$815.85\\-748.66\\\hline\end{array}$ 10. $\begin{array}{r}\$543.72\\-265.65\\\hline\end{array}$

11. $\begin{array}{r}\$436.47\\-357.68\\\hline\end{array}$ 12. $\begin{array}{r}8,628\\-7,645\\\hline\end{array}$ 13. $\begin{array}{r}7,487\\-2,563\\\hline\end{array}$ 14. $\begin{array}{r}\$4,319.53\\-2,428.65\\\hline\end{array}$ 15. $\begin{array}{r}\$47,475.76\\-28,258.68\\\hline\end{array}$

16. $\begin{array}{r}\$14,372.73\\-5,453.80\\\hline\end{array}$ 17. $\begin{array}{r}28,364\\-9,647\\\hline\end{array}$ 18. $\begin{array}{r}35,268\\-27,342\\\hline\end{array}$ 19. $\begin{array}{r}24,834\\-15,748\\\hline\end{array}$ 20. $\begin{array}{r}47,628\\-8,563\\\hline\end{array}$

21. $\begin{array}{r}17,485\\-16,527\\\hline\end{array}$ 22. $\begin{array}{r}403\\-168\\\hline\end{array}$ 23. $\begin{array}{r}104\\-36\\\hline\end{array}$ 24. $\begin{array}{r}2,100\\-834\\\hline\end{array}$

25. $\begin{array}{r}50,304\\-7,347\\\hline\end{array}$ 26. $\begin{array}{r}50,300\\-32,462\\\hline\end{array}$

Exercise B   Find the net amount or balance in each of the following problems:

27.

Gross Pay	$223	15
Deductions	50	17
Net Pay		

28.

Gross Pay	$235	27
Deductions	40	32
Net Pay		

29.

Gross Pay	$265	75
Deductions	46	68
Net Pay		

30.

No. *43*   *$247.50*

Date: _*6/29*_

To: _*A.D. Ross, D.D.S.*_

For: _*Dental work*_

	Dollars	Cents
Balance Brought Forward	463	57
Amount Deposited	—	—
Total	463	57
Amount This Check	247	50
Balance Carried Forward		

31.

No. *63*   *$138.72*

Date: _*8/17*_

To: _*ABC Dept Store*_

For: _*July bill*_

	Dollars	Cents
Balance Brought Forward	472	85
Amount Deposited	—	—
Total	472	85
Amount This Check	138	72
Balance Carried Forward		

32.

No. *72*   *$257.92*

Date: _*11/7*_

To: _*Winters Oil Co*_

For: _*Oil delivery 10/31*_

	Dollars	Cents
Balance Brought Forward	538	67
Amount Deposited	—	—
Total	538	67
Amount This Check	257	92
Balance Carried Forward		

33.

Original Price	$830	00
Sale Price	628	72
Amount of Reduction		

34.

Original Price	$510	00
Amount of Reduction	193	75
Sale Price		

35.

Total Due	$3,428	63
Amount Paid	1,463	74
Balance		

36.

Total Due	$4,216	47
Amount Paid	3,407	63
Balance		

37.

Total Due	$3,427	35
Amount Paid	2,473	64
Balance		

38.

Total Due	$6,527	15
Amount Paid	4,432	63
Balance		

39.

Total Due	$4,263	19
Amount Paid	2,472	49
Balance		

40.

Gross Pay	$263	53
Deductions	47	68
Net Pay		

41.

Gross Pay	$273	45
Deductions	65	53
Net Pay		

Word Problems

PROCEDURE: To solve a word problem:

1. *Understand the problem* by reading it carefully, and answer these questions: What is the required solution? What are the facts of the problem? Will drawing a picture or a chart help? Can you put the facts in one, or more, equation?

2. *Decide on a plan.* What equation(s) will you need that relates the known facts to the desired solution? Set them up.

3. *Carry out the plan.* Solve the equation(s) and get the solution you are looking for.

4. *Check your solution.* Does the solution make sense? Does it satisfy all the conditions of the problem? Are you certain you made no errors?

Solving Equations

Letter symbols called *variables* are sometimes used to represent unknown quantities whose values you have to find.

When these letter symbols include numbers and are separated by an equal sign, they form an *equation* or *formula*.

You can think of an equation as a *balanced scale* with *equal* weights on each side. Whenever you *remove* (or *add*) a weight on *one side* of the scale, you must remove (or add) the same weight on the *other side*. In other words:

Whatever you do on one side of an equation (add, subtract, multiply, or divide), you must also do to the other side.

To solve the equation $n + 7 = 15$, you need to find the *numerical value* of the variable n.

You know that *subtraction* will *undo addition*. Subtracting the *same number* from each side of the equation will leave the *variable* by itself on *one side*, and its *numerical value* on the *other side*, of the equation.

Therefore, to solve the equation $n + 7 = 15$, subtract 7, the number that is added to n, from each side:

$$\begin{array}{r} n + \cancel{7} = 15 \\ -\cancel{7} \quad -7 \\ \hline n = 8 \; Answer \end{array}$$

To *check* the solution, replace n with its *numerical* value, and solve the equation:

$$Check: \quad \begin{array}{l} n + 7 = 15 \\ \hline 8 + 7 = 15 \\ \quad\quad 15 = 15 \checkmark \end{array}$$

 EXAMPLE 7   Carlos has a balance of $846.75 in his checking account. If he writes checks for the following amounts: $158.50, $72.80, and $319.15, what is his new balance?

SOLUTION

Read the problem carefully, *determine* what you need to find, and *outline* the given facts.

GIVEN FACTS: 1. *Starting balance is $846.75.*
2. *Carlos wrote checks for $158.50, $72.80, and $319.15.*

FIND: *The new balance.*
Restate the problem in a single sentence.

The *new balance* will be the *starting balance* ($846.75), *minus* the *sum* of the checks ($158.50 + $72.80 + $319.15).

Translate the sentence into an equation, and *solve* the equation.

NOTE:　Since you want to *add the checks first, place parentheses around these amounts* to indicate that you are to *evaluate the amounts inside the parentheses first.*

New balance = $846.75　−　($158.50 + $72.80 + $319.15)
　　　　　 =　846.75　−　(158.50 + 72.80 + 319.15) *Add first.*
　　　　　 =　846.75　−　　550.45 *Subtract.*
　　　　　 =　296.30
New balance = $296.30

Rules for Order of Operations.
Evaluate expressions in the following order:
1.　Evaluate *inside parentheses () or brackets [] first.*
2.　Next, perform multiplication and division in order, from left to right.
3.　Finally, perform addition and subtraction in order, from left to right.
　　Most calculators are programmed to follow the above *order of operations.* To check the solution to Example 7 using a calculator, enter each number (including decimal points), the symbols ⊞ , ⊟ ,⊠, ⊟ , parentheses as they appear in the problem:（ and）, and finally the = sign.

CHECK:　　846⊡75 ⊟（158⊡50 ⊞ 72⊡80 ⊞ 319⊡15）⊟ 296.30
NOTE:　　Most calculators will display the answer as 296.3 unless the decimal point has been preset at two places.
ANSWER:　Carlos' new balance is $296.30.

Exercise C　Solve the following problems:
42.　A stock bin contained 375 sport shirts. If 128 shirts were removed to the display counter, how many shirts were left in the bin?

43.　Yvonne earned $16,478 this year. Last year her salary was $14,625. How much more did she earn this year?

44.　Albert bought a stereo on the installment plan for $1,128.50. He made a down payment of $475 and made the following monthly payments: May, $123.50; June, $142.85; July, $132.65; and August, $135.62. What is the balance owed on the stereo?

45.　Cindy's monthly take-home pay is $1,375.83. Her monthly expenses are as follows: rent, $465; food, $235.65; telephone, $56.78; travel, $23.50; clothing, $75.90; and other, $325. How much does she have left at the end of the month?

46.　A retailer had the following sales last week: Monday, $4,628.90; Tuesday, $3,426.37; Wednesday, $5,625.22; Thursday, $4,780.65; Friday, $5,816.95; and Saturday, $8,613.50. Merchandise returned by customers for the same days amounted to the following: $368.70, $342.93, $435.65, $415.28, $462.30, and $573.92. What were the net sales for the week?

HORIZONTAL SUBTRACTION

On many business forms where you have to do subtraction, the minuend and subtrahend may be arranged horizontally.

Invoice Number	Amount		Discount		Net Amount	
1075	$565	00	⊟ $84	75	⊜ $480	25
915	118	25	⊟ 17	70	⊜ 100	55

Check:
480 ⊡ 25 ⊞ 84 ⊡ 75 ⊟ 565 ✓
100 ⊡ 55 ⊞ 17 ⊡ 70 ⊟ 118.25 ✓

CHECKING HORIZONTAL SUBTRACTION

As in vertical subtraction, you should *add the difference to the subtrahend* to check your subtraction. This sum should equal the *minuend*.

Many business forms have a built-in check in that you must *subtract horizontally* and *add vertically*.

 Complete the following form:

STEP 1.   Using a calculator, subtract the subtrahend from the minuend on each line, placing the difference in the Net Sales column.

STEP 2.   Using a calculator, add the three columns, putting the totals on the bottom line.

Department	Sales		Returns		Net Sales	
Appliances →	$ 6,824	53	$ 968	65	$5,855	88
Furniture →	11,568	72	1,832	61	9,736	11
Lamps →	4,782	61	648	73	4,133	88
Totals		−		=	Grand Total	

STEP 3.   Subtract horizontally on the bottom line to check your answer.

$$\$23,175.86 - \$3,449.99 = \$19,725.87$$

Department	Sales		Returns		Net Sales		
Appliances	↓ $ 6,824	53	$ 968	65	$ 5,855	88	
Furniture	↓ 11,568	72	1,832	61	9,736	11	
Lamps	↓ 4,782	61	648	73	4,133	88	
Totals	$23,175	86	− $3,449	99	= $19,725	87	Grand Total

$$\$19,725.87 = \$19,725.87$$

EXERCISES

Exercise D   Find the difference in each of the following problems, and check your answers:

47. $87.48 – $62.53

48. $59.37 – $28.50

49. $372.45 – $285.15

50. 28,963 – 14,798

51. $17.83 – $13.89

52. $68.75 – $64.37

53. 132,564 – 129,479

54. 32,865 – 4,973

55. $711.18 – $247.58

56. 3.051 – 2.069

57. 14,678,950 – 2,573,869

58. $16,483.21 – $12,689.45

Exercise E Complete the following forms by subtracting horizontally. Check your answers.

59.

Men's Apparel			Week of 9/2
Item	**Stock**	**Sold**	**On Hand**
Suits	578	138	_____
Slacks	363	156	_____
Sport Jackets	417	125	_____
Coats	378	115	_____
Sweaters	432	163	_____
Sport Shirts	518	235	_____
Totals	−	=	**Grand Total**

60.

Department	Sales		Returns		Net Sales	
Appliances	$ 5,824	58	$1,060	78	_____	___
Furniture	19,568	93	1,038	79	_____	___
Lamps	3,982	67	768	88	_____	___
Totals		−		=		**Grand Total**

61.

Item	Original Price		Amount of Reduction		Sales Price	
A	$9,010	50	$3,420	65	_____	___
B	2,005	00	635	75	_____	___
C	1,300	50	428	64	_____	___
D	4,008	00	1,235	78	_____	___
E	5,040	50	1,836	55	_____	___
F	6,100	00	2,242	45	_____	___
G	3,050	50	943	60	_____	___
Totals		−		=		**Grand Total**

62.

Employee	Gross Pay		Deductions		Net Pay	
Alan, G.	$205	00	$36	48	———	——
Ambrose, L.	240	08	43	69	———	——
Baker, F.	200	15	35	68	———	——
Buntel, M.	190	00	31	38	———	——
Campbell, N.	220	05	36	26	———	——
Chisholm, D.	208	00	38	45	———	——
Fulton, F.	203	00	35	28	———	——
Totals		−		=		

Grand Total

63.

Weekly Net Payroll Week of 10/15						
Employee	Gross Pay		Deductions		Net Pay	
A	$280	75	$73	52	———	——
B	375	50	92	65	———	——
C	347	38	87	35	———	——
D	295	65	81	47	———	——
E	356	90	86	38	———	——
F	325	75	84	10	———	——
G	328	63	84	20	———	——
H	370	95	87	42	———	——
Totals		−		=		

Grand Total

64.

Dep't	Gross Profit		Overhead		Net Profit	
Grocery	$3,462	58	$1,278	63	———	——
Vegetables	2,678	94	963	47	———	——
Dairy	1,875	58	694	78	———	——
Delicatessen	1,684	54	573	38	———	——
Nonfood	1,378	92	485	63	———	——
Meat	2,872	65	984	72	———	——
Totals		−		=		

Grand Total

Unit 3: Multiplication in Business

In business, multiplication is used all the time to calculate inventories, to determine the total cost of a shipment of goods, to figure the hourly wages of an employee, or to calculate the amount of a discount. Multiplication can be thought of as repeated addition. So if you were to sell someone 5 window fans at $49.99 each, the calculation would be 49.99 × 5.

$$
\begin{array}{r}
{\scriptstyle 2\ 4\ 4\ 4} \\
\$49.99 \\
\times\qquad 5 \\
\hline
\$249.95
\end{array}
\quad
\begin{array}{l}
\Leftarrow \textbf{multiplicand} \\
\Leftarrow \textbf{multiplier} \\[4pt]
\Leftarrow \textbf{product}
\end{array}
$$

This is the same as adding $49.99 5 times:

$$
\begin{array}{r}
{\scriptstyle 2\ 4\ 4\ \ 4} \\
\$49.99 \\
\$49.99 \\
\$49.99 \\
\$49.99 \\
+\$49.99 \\
\hline
\$249.95
\end{array}
$$

When multiplying, we count the number of decimal places in the multiplier and the multiplicand (in this case 2). Starting from the right, count left in the product and place the decimal point there. The result is the product.

   Note that as with addition and subtraction, each of the numbers has their own special names. The only name you will probably encounter in business is the answer, the *product*. Note also that we carry, like in addition.

   As a second example, we calculate the cost of a shipment of topcoats. We have just taken delivery on 200 topcoats, each of which will retail for $79.95. What is the value of the shipment?

EXAMPLE 2

$$
200 \times \$79.95
$$

$$
\begin{array}{r}
{\scriptstyle 1\ 1\ \ 1} \\
\$79.95 \\
\times\quad 200 \\
\hline
\$15,990.00
\end{array}
$$

Note that since we had to multiply by 200, we added two zeros to the right end of the answer before we started the multiplication. Always add the same number of zeros as the number of places you are moving (one zero for tens, two for hundreds, and so forth).

Our first two examples had decimals in them. For the third example, look at two-digit whole-number multiplication. How many bed sheets do you have if you have 23 shelves with 13 sheets on each shelf?

$$23 \times 13$$

$$
\begin{array}{r}
23 \\
\times\ 13 \\
\hline
69 \\
230 \\
\hline
299
\end{array}
$$

Note that we do two separate multiplications: for the ones (3) and for the tens (1). We must line up the tens product with the tens place. That is why we have a zero at the right of that product. Then we add up the partial products to get the product. If you have three-digit multiplication, you would do three partial products, as in the next example.

$$537 \times 483$$

$$
\begin{array}{r}
537 \\
\times\ 483 \\
\hline
1\,611 \\
42\,960 \\
214\,800 \\
\hline
259{,}371
\end{array}
$$

With three-digit multiplication, we must line up the tens and hundreds partial products with zeros on the right side, as shown.

CHECKING MULTIPLICATION

As with addition and subtraction, you should always check your multiplication, again using a calculator. To check, divide the product by the multiplier to get the multiplicand. You could instead divide the product by the multiplicand to get the multiplier. If you like using a spreadsheet, you can put in a formula that will multiply one cell by another.

EXERCISES

Exercise A   Solve the following problems, and check your answers:

1. 463 × 28

2. 357 × 54

3. 4,207 × 403

4. 627.24 × 13

5. 728 × 0.35

6. 4,304 × 0.038

7. 386 × 205

8. 4,250.50 × 0.625

9. 1,705.38
   × 0.375

10. 1.030.28
    ×0.0975

11. 3,075.60
    × 0.055

12. 4.206.35
    × 0.0025

13. 5,030.25
    ×0.375

Exercise B   Complete the following forms:

14.

Quantity	Description	Unit Price		Extension	
18 doz.	#1302 Stemware	$18	35/doz.	_____	____
32 sets	#1567 Translucent China	34	62/set	_____	____
27 sets	#158 Bavarian China	47	80/set	_____	____
18 sets	#758 Coffee Sets	12	73/set	_____	____
			Total		

15.

ORDERED BY: A. Rivera	SHIP: Before 9/15	SHIP VIA: Truck		TERMS: Usual	
Quantity	Description	Unit Price		Extension	
24 sets	Bone China, Service for 8, #1728	$19	75	——	—
18 sets	Bone China, Service for 12, #2071	34	50	——	—
36 sets	English Tea Sets, #509	12	45	——	—
14 sets	Salad Sets, #1249	9	85	——	—
24 sets	Tea and Dessert Sets, #1415	13	95	——	—
			Total		

16.

SALESPERSON: G. Santini	DATE SHIPPED: 7/15	SHIPPED VIA: Freight	TERMS: 2/20, n 60	PURCHASE ORDER NO.: 4360	
Quantity	Description	Unit Price		Amount	
28 boxes	Floor Tiles #1106	$8	15	——	—
150 boxes	Floor Tiles #1108	6	25	——	—
248 boxes	Bathroom Tiles, Yellow	5	78	——	—
175 boxes	Bathroom Tiles, Blue	5	49	——	—
375 gal.	Tile Adhesive #708	2	86	——	—
225 gal.	Tile Adhesive #710	3	25	——	—
			Total		

17.

INVENTORY FORM

Stock No.	Quantity	Description	Unit Price		Extension	
307	63	Sport Jackets	$ 27	85	_____	___
461	123	Overcoats	63	35	_____	___
107	235	Flared Slacks	12	95	_____	___
423	164	Sport Jackets	43	75	_____	___
613	417	Regular Slacks	15	65	_____	___
707	315	Raincoats	27	90	_____	___
321	128	Leather Belts	5	65	_____	___
				Total		

18.

SOLD TO ___Handy Andy Hardware___ INVOICE NO. ___7342___

___2316 Union Street___ DATE ___May 14___ 20 ___ -- ___

___Cleveland, Ohio 44117___ OUR ORDER NO. ___3620___

CUSTOMER'S ORDER NO. ___572___

TERMS ___3/15, net 45___ SHIPPED VIA ___Truck___

Quantity	Description	Unit Price		Total Amount	
245 gal.	Interior Latex, Flat, White	$ 7	85	_____	___
175 gal.	Interior Latex, Flat, Pink	9	85	_____	___
250 gal.	Interior Latex, Flat, Royal Blue	11	65	_____	___
175 gal.	Interior Latex, Flat, Moss Green	8	50	_____	___
235 gal.	Interior Latex, Flat, Lemon Sage	6	95	_____	___
265 gal.	Interior Latex, Flat, Burnt Orange	9	75	_____	___
			Total		

Word Problems

As with addition, the key phrases indicating multiplication are:

"What is the total?"

"Find the cost of. . . ."

Use multiplication if the problem gives you a unit price and you have to find the cost of a number of units, if you are asked to change a larger unit of measure to a smaller unit, or if you are given a salary rate and have to find a weekly, monthly, or yearly salary.

Many business problems have more than one part; you must add the total cost of one item to the total cost of another (see Exercise 22).

Exercise C   Solve the following problems:

19. Ace Used Cars sells a car for $350 down and 36 monthly payments of $79. What is the total cost of the car?

SAMPLE SOLUTION

Solve the problem, and check the answer on a calculator.

First, outline the problem:

GIVEN FACTS: 1. $350 is the down payment.

2.   There are also 36 payments of $79 each.

FIND: The total cost of the car.

The *total* cost will be the down payment, $350, *plus* 36 payments of $79 each.

$$\text{Total cost} = \$350 + (36 \times \$79)$$

Evaluate the inside parentheses first.

Total cost = $350 + (36 × $79) Multiply first.

= $350 + $2,844 Add.

= $3,194

```
   36      2,844
 ×79      + 350
  324      3,194
 252
2,844
```

CHECK: 350 ⊞ ⟮ 36 ⊠ 79 ⟯ ⊟ 3194 ✓

ANSWER: Total cost of the used car is $3,194.

20. Marking pens sell at $4.95 per dozen. How much will 124 dozen cost?

21. The Mutual Insurance Company bought 18 four-drawer steel filing cabinets. The cost of each cabinet was $123.75. Find the total cost of the cabinets.

22. What will be your gross pay for 51 hours of work in 1 week if you get paid $5 per hour and receive time and a half for work beyond 40 hours? (HINT: First find the cost of labor for 40 hours by multiplying 40 × $5. Then, find the cost of overtime by multiplying 11 hours by $7.50. Finally, find the total cost by adding the regular and the overtime pay.)

23. The cost analysis of producing a 19" color TV is as follows: parts, $67.35; labor, $43.75; overhead, $19.65; cost of selling, $22.85; advertising, $12.50; and shipping, $5.95. What is the total cost of producing 125 TV sets?

ROUNDING NUMBERS

You round numbers when you want to use an *approximate* value of a number, instead of its *exact value*.

• If, for a given year, the total sales of a company amounted to $2,845,698, the company may want to list the total sales as "close to 3 million dollars."

The *exact amount*, $2,845,698, is *closer* to $3,000,000 than it is to $2,000,000. *Rounded to the nearest million*:

$$\$2,845,698 \approx \$3,000,000$$

The symbol ≈ means "is approximately equal to."

• If the solution to a problem involving *dollars and cents* is $19.3843, the number 3843 should be rounded to the nearest penny since fractions of a penny cannot be collected or paid.

The *exact amount*, $19.3843, is *closer* to $19.38 than it is to $19.39. *Rounded to the nearest penny*:

$$\$19.3843 = \$19.38$$

REVIEW OF PLACE-VALUE COLUMNS AND NAMES AND VALUES OF NUMBERS
Place-Value Columns

In a *mixed decimal number*, the *decimal point* separates the *whole-number* part from the *decimal part*.

For the number 628.247, *the place values* of the columns are as follows:

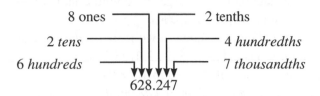

The number 628.247 is read as follows:

Six hundred twenty-eight and two hundred forty-seven thousandths

Note that the *decimal point* translates to the word *and*.

- With *whole numbers*, moving *left from the decimal point*, each column is *multiplied by 10*, and is *10 times as large* as the column to its *right*.

- With *decimal numbers*, moving *right from the decimal point*, each column is *divided by 10*, and is $\frac{1}{10}$ the value of the column to its *left*.

The accompanying table gives the names and values of whole numbers and decimal numbers.

Place-Value Columns of Whole Numbers and Decimal Numbers

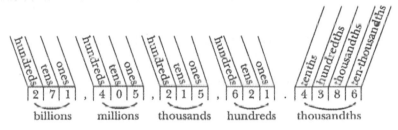

Names and Values of Numbers

- *Whole Numbers*
  A *three-digit* number names a value in *hundreds* (tens, ones).
  A *six-digit* number names a number in *thousands* (hundreds, tens, ones).
  A *nine-digit* number names a value in *millions* (thousands, hundreds, tens, ones).
  A *twelve-digit* number names a value in *billions* (millions, thousands, hundreds, tens, ones).

- *Decimal Numbers*
  A *one-place* decimal number names a value in *tenths*.
    $0.5 = 5$ tenths $\left(\frac{5}{10}\right)$; $0.9 = 9$ tenths $\left(\frac{9}{10}\right)$
      (1-place decimal: 1 zero in the number 10)
  A *two-place* decimal number names a value in *hundredths*.
    $0.05 = 5$ hundredths $\left(\frac{5}{100}\right)$; $0.25 = 25$ hundredths $\left(\frac{25}{100}\right)$
      (2-place decimal: 2 zeros in 100)

A *three-place* decimal number names a value in *thousandths*.

0.005 = 5 thousandths $\left(\frac{5}{1,000}\right)$; 0.125 = 125 thousandths $\left(\frac{125}{1,000}\right)$
(3-place decimal: 3 zeros in 1,000)

A *four-place* decimal number names a value in *ten-thousandths*.

0.0005 = 5 ten-thousandths $\left(\frac{5}{10,000}\right)$; 0.0125 = 125 ten-thousandths
$\left(\frac{125}{10,000}\right)$ (4-place decimal: 4 zeros in 10,000)

As you continue moving *right from the decimal point*, each column is *divided by 10*, and is $\frac{1}{10}$ *the value* of the column to its *left*.

REVIEW OF ROUNDING SKILLS

Rounding Decimal Numbers

Underline the digit to be rounded. (In rounding to the *nearest penny*, underline the hundredths place.)

1. If the digit to the right is 5 or more, *add 1* to the *underlined digit*. If the digit to the right is less than 5, do *not* add 1.

2. Drop all digits to the right of the underlined digit.

Rounding Whole Numbers

Underline the digit to be rounded.

1. If the digit to the right is 5 or more, *add 1* to the *underlined digit*. If the digit is less than 5, do *not* add 1.

2. Replace all digits to the right of the underlined digit with zeros.

 Round $287.8967 to the nearest *penny*.

SOLUTION

Underline the 9, the *hundredths* place, and check the digit to the right.

$287.8<u>9</u>67   Since the digit to the right of 9 is *more* than 5, *add* 1 to the 9, and drop all digits to the right of the rounded digit.

ANSWER: $287.8967 = $287.90 rounded to the nearest penny

 Round $257,489 to the nearest *thousand*.

SOLUTION

Underline the 7, the *thousands* place, and check the digit to the right.

$25<u>7</u>,489   Since 4 is *less* than 5, do *not* add 1 to the 7. Replace all digits to the right of the underlined digit with zeros.

ANSWER: $257,489 = $257,000 rounded to the nearest thousand.

EXAMPLE 7   What is the place value of each digit in $1,234,567.89?

- The 1 is in the *millions* place and means 1,000,000.

- The 2 is in the *hundred-thousands* place and means 200,000 (2 × 100,000).

- The 3 is in the *ten-thousands* place and means 30,000 (3 × 10,000).

- The 4 is in the *thousands* place and means 4,000 (4 × 1,000).

- The 5 is in the *hundreds* place and means 500 (5 × 100).

- The 6 is in the *tens* place and means 60 (6 × 10).

- The 7 is in the *ones* place and means 7 (7 × 1).

- The 8 is in the *tenths* place and means 0.80 (8 × 0.10).

- The 9 is in the hundredths place and mean 0.09 (9 × 0.01).

EXERCISES

Exercise D   Multiply, and round each answer to the indicated place.

24. $235.73
    × 0.36
    (cent)

25. $405.23
    × 0.025
    (dollar)

26. $365.98
    × 0.06
    (cent)

27. $938.75
    × 0.025
    (dollar)

28. $1,478.62
    × 0.075
    (cent)

29. 2,460.50
    × 1.405
    (thousand)

30. $3,050.75
    × 0.005
    (cent)

31. $4,367.80
    × 0.0525
    (cent)

32. $5,362.78
    × 0.008
    (cent)

33. $4,378.92
    × 2.05
    (cent)

Exercise E   Complete these forms, rounding each number to the nearest *cent*.

34.

Quantity	Description	Unit Price		Extension	
43.25 gal.	Alcohol, Wood	$ 6	75	_____	___
57.5 gal.	Paint Thinner	7	35	_____	___
85.35 gal.	Solvent #5	8	45	_____	___
73.65 gal.	Solvent #8	9	65	_____	___
65.65 gal.	Alcohol, Denatured	11	85	_____	___
128.80 gal.	Clear Lacquer	6	35	_____	___
			Total		

35.

Quantity	Description	Unit	Unit Price		Extension	
5.56 lb.	Potassium	Pound	$17	50	_____	___
7.38 kg	Magnesium	Kilogram	14	75	_____	___
7.53 lb.	Sulfur	Pound	22	47	_____	___
				Total		

36.

Quantity	Description	Unit Price		Extension	
23.8 tons	Wheat	$68	25	_____	___
16.43 tons	Rice	83	37	_____	___
18.63 tons	Soya Bean	72	46	_____	___
20.31 tons	Corn	57	19	_____	___
17.07 tons	Oat	63	85	_____	___
			Total		

Round each numbers to the place values indicated.

37.

Amount	$1	$100	$1,000	$10,000	$100,000
$ 356,247 93	_____	_____	_____	_____	_____
474,563 42	_____	_____	_____	_____	_____
1,263,437 59	_____	_____	_____	_____	_____
835,726 40	_____	_____	_____	_____	_____
452,374 75	_____	_____	_____	_____	_____
168,935 15	_____	_____	_____	_____	_____
244,568 53	_____	_____	_____	_____	_____

Word Problems

Exercise F Solve the following problems:

38. Joise bought three blouses at $32.95 each, two skirts at $45.95 each, and a jacket for $65.95. If the sales tax is 6%, what is the total cost of her purchase? (6% = 0.06)

39. An appliance manufacturer sells a refrigerator at a list price of $685.75 less a trade discount of 40%. What is the net price of the refrigerator? (40% − 0.40)

DEVELOPING SPEED IN MULTIPLICATION

In all businesses, speed as well as accuracy is needed in day-to-day calculations. A person with these two abilities will gain a positive reputation among supervisors and colleagues.

MULTIPLICATION BY A POWER OF 10

A useful shortcut involves multiplying by a power of 10, such as 100 or 1,000. This type of multiplication can usually be done without written calculations.

PROCEDURE To multiply by a power of 10:

1. When the multiplicand is a whole number, *multiply the numbers that are not zero and attach the same number of zeros to the product as there are zeros in the multiplicand and multiplier*. What you are really doing is moving the decimal point one place to the right for each zero in the multiplicand and multiplier.

2. When the multiplicand is a decimal or mixed decimal, *move the decimal point in the multiplicand to the right as many places as there are zeros in the multiplier*.

Whole numbers

354×10	$= 354 \times 1 = 3{,}540$	Attach 1 zero.
354×100	$= 354 \times 1 = 35{,}400$	Attach 2 zeros.
$354 \times 1{,}000$	$= 354 \times 1 = 354{,}000$	Attach 3 zeros.
25×30	$= 25 \times 3 = 750$	Attach 1 zero.
420×20	$= 42 \times 2 = 8{,}400$	Attach 2 zeros.
130×300	$= 13 \times 3 = 39{,}000$	Attach 3 zeros.

Decimals and mixed decimals

0.354×10	$= 0.354$	$= 3.54$	Move the decimal point 1 place to the right.
3.54×100	$= 3.54$	$= 354$	Move the decimal point 2 places to the right.
$3.54 \times 1{,}000$	$= 3.540$	$= 3{,}540$	Move the decimal point 3 places to the right.
0.125×30	$= 0.375$	$= 3.75$	Move the decimal point 1 place to the right.
1.25×300	$= 3.75$	$= 375$	Move the decimal point 2 places to the right.
$12.5 \times 3{,}000$	$= 37.500$	$= 37{,}500$	Move the decimal point 3 places to the right.

MULTIPLICATION BY A DECIMAL THAT IS A POWER OF 10

This method can also be used with decimals that are powers of 10, such as 0.1, 0.01, 0.001, and 0.0001.

PROCEDURE   To multiply a number by a decimal that is a power of 10, move the decimal point in the multiplicand as many places to the *left* as there are decimal digits in the multiplier, inserting zeros if necessary.

NOTE: When you multiply by a decimal of 10, or a power of 10, you are *decreasing the product* by 10 or powers of 10. Therefore, *move the decimal point to the left.*

123.45×0.1	$= 12.345$	Move the decimal point 1 place to the left.
123.45×0.01	$= 1.2345$	Move the decimal point 2 places to the left.
123.45×0.001	$= 0.12345$	Move the decimal point 3 places to the left.
123.45×0.0001	$= 0.012345$	Move the decimal point 4 places to the left.

CAUTION: When using this shortcut, disregard any zeros at the end of the multiplier. A decimal such as 0.10 or 0.100 will move the decimal point *one place* to the left in the multiplicand ($0.100 = 0.10 = 0.1$).

EXERCISES

Exercise G Find the following products without using written calculations:

40. 427 × 0.01   41. 0.534 × 100   42. $5.15 × 1,000   43. $28 × 100

44. $0.09 × 100   45. $12.50 × 0.20   46. $30.15 × 300   47. 300 × $10.50

48. 20% × 450   49. 43.50 × 200   50. 1.50 × 3,000   51. 13.50 × 200

Exercise H Complete the following forms, doing the multiplication without using written calculations:

52.

Quantity	Description	Unit Price		Amount	
200 boxes	#75 Screwdrivers	$ 3	75	_____	___
20 cases	#19 Nails	2	50	_____	___
1,000 boxes	#22 Nails		53	_____	___
3,000 boxes	#15 Nails	1	30	_____	___
200 boxes	6" Bench Wire	12	41	_____	___
			Total		

53.

Quantity	Description	Unit Price		Amount	
20 boxes	#1207 Full-fashioned Sweaters	$47	75	_____	___
100 boxes	#157 Long-sleeve Blouses	24	75	_____	___
85	#307 Short Jackets	10	00	_____	___
32	#461 Full-length Jackets	30	00	_____	___
			Total		

54.

Quantity	Description	Unit Price		Amount	
200 doz.	#2 Pencils	$ 1	50	_____	___
60 gross	#7 Erasers	4	00	_____	___
180 doz.	#105 Ball-point Pens	20	00	_____	___
2,000	12″ Rulers		53	_____	___
5,000	8″ Rulers		30	_____	___
			Total		

Unit 4: Division in Business

In business, division is used to calculate how many of a shipment of goods will be divided among a number of stores, to determine the unit price of each item in that shipment, or in an installment plan to calculate how much money must be paid for each of the installments. All of these, and many more calculations, involve division. Division is the inverse of multiplication, just as subtraction is the inverse if addition. If you multiply 5×9, you get 45. So if you divide 45 by 9, you get 5.

$$45 \div 9 = 5 \quad \text{or} \quad 9\overline{)45}^{\,5} \quad \text{or} \quad 45/9 = 5 \quad \text{or} \quad \frac{45}{9} = 5$$

All of these are ways of showing division. You probably will not see the first two very often. In this example, 45 is the dividend, 9 is the divisor, and 5 is the quotient.

LONG DIVISION

Although you probably will not use this method too often since it is considerably longer than the other three operations, we will go over it here. A shipment of one gross (144) of quart bottles of dishwashing liquid came in, and we need to place them onto shelves. Each shelf will hold 9 bottles. How many shelves do we need?

$$9\overline{)144}$$

SOLUTION

Just like in addition, subtraction, and multiplication, we divide one place at a time. Unlike the other three, we start on the left. We ask, how many 9s go into 1 (hundred)? The answer is none. So we look at the next number. How many 9s go into 14 (tens)?

The answer is 1. So we put the 1 up top, lined up with the 14. Then we multiply 1 by 9 and line that up below the 14. Then we subtract 9 from 14 to get 5, like this:

$$
\begin{array}{r}
1 \\
9{\overline{\smash{)}144}} \\
\underline{9} \\
54
\end{array}
$$

We drop the 4 down next to the 5 and ask, how many 9s go into 54? The answer is 6. So we put a 6 up top, lined up with the 4. Then we multiply 9 by 6, get 54, and put that under the 54. Then we subtract, getting 0, like this:

$$
\begin{array}{r}
16 \\
9{\overline{\smash{)}144}} \\
\underline{9} \\
54 \\
\underline{54} \\
0
\end{array}
$$

This is an easy one because it has an answer with no remainder. We will need 16 shelves for the 144 bottles of dishwashing liquid, and each shelf will hold 9 bottles. If our shelves held 10 bottles, what happens then? Well, here is the division problem again but with 10 as the divisor.

$$
\begin{array}{r}
14 \\
10{\overline{\smash{)}144}} \\
\underline{10} \\
44 \\
\underline{40} \\
4
\end{array}
$$

You see that the quotient is 14, with a remainder of 4. So we need 14 shelves to hold 140 bottles. We will also need another shelf to hold the remaining 4 bottles.

Now, let's do an example that has decimal places in it. We have a consignment of 48 gallon bottles of antifreeze. They cost $175.00. How much does each bottle cost? We set up the division problem.

EXAMPLE 3

$$
48{\overline{\smash{)}175.00}}
$$

Note that we put a decimal point up top, lined up with the other. So we ask, how many 48s can go into 17? The answer is none. So we ask how many 48s go into 175? The answer is 3. So we put 3 lined up with 175, multiply 3 by 48, get 144, put it below the 175, and subtract. Then we being down the 0.

$$
\begin{array}{r}
3. \\
48{\overline{\smash{)}175.00}} \\
\underline{144} \\
31.0
\end{array}
$$

Now we are on the right side of the decimal place. We now ask, how many 48s go into 310? The answer is 6. So we multiply 48 by 6, get 288, subtract that from 310, get 22, and bring down the last 0.

$$
\begin{array}{r}
3.6 \\
48\overline{)175.00} \\
\underline{144} \\
31.0 \\
\underline{28\ 8} \\
2\ 20
\end{array}
$$

Next we ask, how many 48s go into 220? The answer is 4. So we multiply 4 by 48 and get 192. We then subtract this from 220 and get 28.

$$
\begin{array}{r}
3.64 \\
48\overline{)157.00} \\
\underline{144} \\
31.0 \\
\underline{28\ 8} \\
2\ 20 \\
\underline{1\ 92} \\
28
\end{array}
$$

Now do one more division. Add a 0 to the 28 to get 280. Then ask, how many 48s go into 280? The answer is 5. So, 48 times 5 is 240, subtract that from 280, and get 40.

$$
\begin{array}{r}
3.645 \\
48\overline{)175.00} \\
\underline{144} \\
31.0 \\
\underline{28\ 8} \\
2\ 20 \\
\underline{1\ 92} \\
280 \\
\underline{240} \\
40
\end{array}
$$

We are at the thousandths place, and we need only hundredths to get the price to the penny. So we round up and determine the price of each bottle of antifreeze to be $3.65. You should divide one place beyond the place you want and then round off.

CHECKING DIVISION

As with addition, subtraction, and multiplication, you should check your division. It is a snap using a calculator. To check, multiply the divisor by the quotient to get the dividend. If you like using a spreadsheet, you can put in a formula that will divide one cell by another.

EXERCISES

Exercise A   Find the quotient in each of the following problems:

1. $0.012\overline{)14.232}$
2. $15\overline{)12.405}$
3. $8.4\overline{)128.624}$
4. $640\overline{)4,480}$

5. $0.358\overline{)223.392}$
6. $0.342\overline{)526.37}$
7. $41.7\overline{)9,678.57}$
8. $305\overline{)65,885}$

9. $420\overline{)639.453}$
10. $280\overline{)481.880}$
11. $415\overline{)88.435}$

12. $0.149\overline{)538,620}$
13. $0.238\overline{)74,763.5}$
14. $24.67\overline{)0.0002467}$

Exercise B   Complete the following forms. Remember to round answers where necessary.

15.

Item	Bulk Unit	Bulk Price		Unit Price	
#2 Pencils	doz. in box	$	72	____	____
Memo Pads	36 in box	47	16	____	____
Typ. Erasers	24 in box	1	92	____	____
Typ. Ribbons	18 in box	11	70	____	____
#10 Envelopes	8 boxes	73	36	____	____
#6 Envelopes	12 boxes	54	60	____	____

16.

Item	Total Amount		Price per Yard		Number of Yards
Carpeting, Blue	$ 850	50	$ 6	75	_____
Carpeting, Plum	983	40	7	45	_____
Carpeting, Gold	1,287	00	9	75	_____
Carpeting, Brick Red	1,502	20	12	95	_____
Carpeting, Royal Blue	1,718	75	13	75	_____

17.

Quantity	Item	Description	Total Cost		Unit	Unit Price	
247 sq. yd.	Carpeting	#327 Shag	$1,482	00	Sq. yd.	___	___
315 boxes	Tiles	#10-5 Embossed	1,575	00	Box	___	___
80 yd.	Fabric	#48 Velvet	400	00	Yard	___	___
215 rolls	Wallpaper	#37-5 Textured	1,505	00	Roll	___	___
235 gal.	Paint	#511 Flat	1,175	00	Gallon	___	___

18.

Amount of Loan		Number of Payments	Amount of Each Payment	
$2,475	65	24	_____	___
6,570	90	36	_____	___
1,258	75	12	_____	___
8,562	80	36	_____	___
9,675	65	48	_____	___
4,860	45	36	_____	___
7,637	85	48	_____	___
3,578	80	24	_____	___
5,685	75	36	_____	___
1,325	80	12	_____	___

19.

Item	Number of Pounds	Total Cost		Cost per Pound	
1	258	$ 315	08	____	____
2	337	1,120	00	____	____
3	139	495	63	____	____
4	453	1,305	72	____	____
5	532	1,620	05	____	____
6	263	1,432	35	____	____
7	385	1,724	50	____	____
8	657	2,128	72	____	____
9	315	1,015	62	____	____
10	225	1,420	68	____	____

20.

Item	Number of Yards	Total Cost		Cost per Yard	
1	85	$ 347	60	____	____
2	105	529	55	____	____
3	215	767	90	____	____
4	325	943	35	____	____
5	467	1,415	70	____	____
6	235	3,267	90	____	____
7	435	4,585	38	____	____
8	355	5,628	15	____	____
9	525	7,293	80	____	____
10	437	6,585	95	____	____

Word Problems

Key phrases indicating division include:

"What is the unit cost?"
"How much is each item?"
"What is the monthly cost?"
"How many months will be needed?"
"What is the . . . per . . .?"

Use division if you are given a large total and are asked to break it down into smaller parts (months, per piece, etc.).

Exercise C   Solve the following problems. Round answers where necessary.

21. The Adler Manufacturing Company ordered 450 yards of velvet at a total cost of $4,950. Find the cost per yard.

22. If 123 gallons of paint cost $615, what is the cost of 1 gallon?

23. If four dozen sport shirts cost $348, how much does one shirt cost?

24. A skirt manufacturer shipped 75 skirts to a customer for a total cost of $1,725. What is the cost of each skirt?

DEVELOPING SPEED IN DIVISION

There are shortcuts in division that can speed the task and improve the accuracy. As noted previously, both speed and accuracy are qualities of major importance to persons in business.

The most useful shortcut in division involves dividing by 10 or a power of 10, that is, 100, 1,000, and so on. This shortcut is also useful when there are end zeros in both the dividend and the divisor.

DIVISION BY A POWER OF 10

PROCEDURE: When dividing by 10 or a power of 10, move the decimal point in the dividend *one place* to the *left* for each zero in the divisor. For example:

$$23. \div 10 \quad = 2.3$$
$$23. \div 100 \quad = 0.23$$
$$023. \div 1,000 = 0.023$$

NOTE: In the example 23 ÷ 1,000, a zero is placed in front of the dividend because the decimal point was moved *three* places and the original dividend consisted of only *two* digits. The zero serves as a *placeholder* for the *tenths* column.

Here are three other examples:

$$\$13.50 \div 10 \quad = \$1.35$$
$$\$25.82 \div 100 \quad = \$0.2582 \text{ or } \$0.26$$
$$\$129.30 \div 1,000 = \$0.1293 \text{ or } \$0.13$$

NOTE: In business computations, you will find that many items are priced in lots of 100 or 1,000, or of a short ton, which equals 2,000 pounds. When calculating totals, remember that *cwt.* means *100 pounds*, C means *100 units,* and *M* means *1,000 units,* and use the power-of-10 shortcut to find the solution.

Find the cost of 657 pounds of rice at $16.50 per cwt.

STEP 1.   657 ÷ 100 = 6.57   Find the number of hundreds
                                          (cwt.) in 657 by moving the deci-
                                          mal point in 657 two places to the
                                          left.

STEP 2.   6.57 × $16.50 = $108.4050   Multiply 6.57, the number of
                      = $108.41 *Answer* hundreds in 657, by $16.50,
                                          the price per cwt.

CHECK: 108 ⊡ 41 ⊟ 16 ⊡ 5 ⊟ 6.57 ✓

Find the cost of 275 pens at $12.50 per 100 pens.

STEP 1.   275 ÷ 100 = 2.75   Find the number of hundreds in
                                          275 by moving the decimal point
                                          in 275 two places to the left.

STEP 2.   2.75 × $12.50 = $34.3750   Multiply 2.75, the number of
                      − $34.38 *Answer*   hundreds in 275, by $12.50, the
                                          price per 100.

CHECK: 34 ⊡ 375 ⊟ 12 ⊡ 5 ⊟ 2.75 ✓

DIVISION WITH END ZEROS IN DIVIDEND AND DIVISOR

When a division problem has end zeros in both dividend and divisor, you may cross
out an *equal number* of end zeros in both *dividend* and *divisor*. What you are really
doing is dividing both dividend and divisor by 10 or a power of 10.

Divide 2,400 by 800.

$$2{,}4\cancel{0}\cancel{0} \div 8\cancel{0}\cancel{0} = 3 \text{ } Answer \qquad 8\cancel{0}\cancel{0}\overline{)2\cancel{40}\cancel{0}}^{\cancel{3}}$$

$$\underline{24}$$

CHECK: 800 × 3 = 2,400 ✓

Divide 7,500 by 250.

$$7{,}50\cancel{0} \div 25\cancel{0} = 30 \text{ } Answer \qquad 25\cancel{0}\overline{)750\cancel{0}}^{30}$$

$$\underline{75}\downarrow$$
$$0$$

CHECK: 250 × 30 = 7,500

Find the cost of 4,600 pounds of coal at $36.50 per ton.

STEP 1. $4,6\cancel{0}\cancel{0} \div 2,0\cancel{0}\cancel{0} = 46 \div 20 = 4.6 \div 2 = 2.3$ Find the number of tons in 4,600 pounds by dividing 4,600 by 2,000.

STEP 2. $2.3 \times \$36.50 = \$83.95$ *Answer*   Multiply that number by $36.50, the price per ton.

Step 1 can sometimes be done without written calculation by using the following method:

PROCEDURE: When dividing by multiples of powers of 10:

1.   Move the decimal point to the left as many places as there are zeros in the divisor.
2.   Divide the new dividend by the left-hand digit in the original divisor.

In Example 5, the divisor 2,000 has three zeros. Therefore, the dividend 4,600 becomes 4.6, and the problem becomes 4.6 ÷ 2.

DIVISION BY A DECIMAL THAT IS A POWER OF 10

PROCEDURE: Dividing by a decimal of 10 or power of 10 will *increase* the quotient by 10, 100, 1,000, and so forth. Therefore, move the decimal point in the *dividend* to the *right* as many places as there are decimal places in the *divisor*.

For example: $4.32 \div 1$ $= 4.32$
$4.32 \div 0.1$ $= 43.2$

$$0.1\overline{)4.3\underset{\frown}{2}}^{\,43.2}$$

$4.3\underset{\frown}{2} \div 0.01$ $= 432$
$4.3\underset{\frown}{2} \div 0.001$ $= 4,320$

Note the insertion of the zero in 4,320, thus maintaining the position of the decimal point.

Divide 1,350 by 0.02.

$$1,350 \div 0.02 = 135,000 \div 2$$
$$= 67,500 \text{ *Answer*}$$

EXERCISES

Exercise D Using the shortcuts shown in this unit, and without using written calculations, find the answer for each of the following problems:

25. $23.45 ÷ 10   26. $56.47 ÷ 1,000   27. 4,210 ÷ 1,000   28. $1.86 ÷ 100

29. 124,000 ÷ 1,500   30. 0.63 ÷ 10   31. 35 ÷ 100   32. $3,240 ÷ 1,000

33. $123.15 ÷ 1,000   34. $2,500 ÷ 500   35. $360 ÷ 0.2   36. $396.45 ÷ 0.03

37. $2,550 ÷ 0.05

Exercise E Using the shortcuts shown in this unit, in each of the following problems find the costs of the items listed:

38.

Number of Items	Description	Price per C		Amount	
475	Pens	$ 9	50	———	—
350	Notebooks	25	80	———	—
1,250	Erasers	2	35	———	—
1,465	Pencils	3	42	———	—
970	Pens	6	25	———	—

39.

Total Number of Pounds	Description	Price per Ton		Amount	
8,400	Rice	$375	60	———	—
12,648	Wheat	237	75	———	—
7,820	Barley	347	50	———	—
5,695	Peanuts	237	95	———	—
10,150	Walnuts	425	50	———	—

Word Problems

Exercise F   Solve the following problems, using the shortcut methods:

40. A retailer bought 10 watches for $475. How much did one watch cost?

41. Address labels are sold at $2.75 per M. How much will 6,750 labels cost?

Unit 5: Fractions in Business

Fractions can be found in every aspect of business. Coffee and tea are sold in pounds and fractions of pounds. A large shipment of anything (jackets, for instance) can be broken up into fractions to go to several stores in a franchise. Many items are sold in dozens or parts of a dozen. Very often hourly wages are calculated to half an hour (and a quarter of an hour, sometimes), and overtime is usually time and a half (1½).

REVIEW OF COMMON FRACTIONS

A fraction is a part of a whole. A common fraction looks like this:

$$\frac{3}{4} \begin{array}{l} \Leftarrow \text{ numerator} \\ \Leftarrow \text{ denominator} \end{array}$$

As you can see, each part of the fraction has a special name. The bottom number, the *denominator*, tells how many parts the whole has been broken into, in this case 4. So the denominator is a divisor. The top number, the *numerator*, tells how many parts we have, in this case 3. So the numerator is a multiplier. A fraction can be thought of as a division exercise. If you divide 3 by 4, you will get 0.75. In just the same way, ¾ of a dollar is 75 cents. Thus, all fractions are another way of looking at a decimal. If you divide the numerator of any fraction by the denominator, you will get a decimal.

FRACTIONS ARE DECIMALS ARE PERCENTS

There is a third very common way of looking at fractions, and that is percents. Percent is simply "part of a hundred." So the example of ¾ is 0.75 as a decimal and also is 75% as a percent. Thus, all three of these (fractions, decimals, and percents) are simply ways of looking at part of a whole. We will do more decimals and percents later in the book.

IMPROPER FRACTIONS

Another type of fraction occurs when you have more than a whole. Consider eggs. These days you can buy them in a package of 18. Usually eggs come as a dozen, so if we were to show this larger package as a fraction out of a dozen, it would be:

$$\frac{18}{12}$$

We call this fraction an *improper fraction* because the numerator is larger than the denominator. We could separate the fraction by taking out a whole dozen, and we would have half a dozen left. So this fraction is 1½ dozen.

MIXED NUMBERS

A mixed number is another way of naming a fraction. It has a whole-number portion and a fraction portion:

$$2\frac{3}{4} \qquad 5\frac{7}{8} \qquad 12\frac{2}{3}$$

All of these are mixed numbers because they all have a whole-number part and a fractional part. Changing a mixed number to an improper fraction is very easy. Simply multiply the whole number by the denominator part of the fraction and then add this to the numerator part. Now let's do that for the three mixed numbers listed:

 EXAMPLE 1 $2\frac{3}{4}$ gives us $2 \times 4 = 8$ and $8 + 3 = 11$

$$2\frac{3}{4} = \frac{11}{4}$$

EXAMPLE 2 $5\frac{7}{8}$ gives us $5 \times 8 = 40$ and $40 + 7 - 47$

$$5\frac{7}{8} = \frac{47}{8}$$

EXAMPLE 3 $12\frac{2}{3}$ gives us $12 \times 3 = 36$ and $36 + 2 = 38$

$$12\frac{2}{3} = \frac{38}{3}$$

REDUCING FRACTIONS

Consider these two fractions:

$$\frac{6}{10} \qquad \frac{3}{5}$$

These are considered *equivalent fractions* because they name the same amount. In the first fraction, if we divide the 6 by 2 and the 10 by 2, we will get 3/5. What we have just done is *reduce* the fraction. Note that we divided both the numerator and denominator by the *same* number, 2. Here is another example:

$$\frac{24}{48}$$

If you divide 24 by 2 and 48 by 2, you will get:

$$\frac{12}{24}$$

Then divide the 12 by 2 and the 24 by 2 to get:

$$\frac{6}{12}$$

Next, divide the 6 by 2 and the 12 by 2 and get:

$$\frac{3}{6}$$

Finally, divide the 6 by 3 and the 12 by 3 to get:

$$\frac{1}{2}$$

Did you notice that you could have divided both the numerator (24) and the denominator (48) of the original fraction, 24/48, by 24 and gotten the same fraction in the end? This means that,

$$\frac{24}{48} \quad \text{and} \quad \frac{1}{2} \quad \text{are equivalent fractions.}$$

To recap, we reduce a fraction by dividing both the numerator and the denominator by the same number.

Going the other way, to raise a fraction to a higher equivalent, simply multiply both the numerator and the denominator by the same number. Now raise

$$\frac{3}{7} \quad \text{to an equivalent fraction.}$$

First multiply both numerator and denominator by 6: $3 \times 6 = 18$ and $7 \times 6 = 42$. We see that

$$\frac{3}{7} \text{ is equivalent to } \frac{18}{42}.$$

Try another example.

 EXAMPLE 4   Change $\dfrac{7}{8}$ to its equivalent by multiplying 5.

$$7 \times 5 = 35 \quad \text{and} \quad 8 \times 5 = 40$$

$$\frac{7}{8} \text{ is equivalent to } \frac{35}{40}$$

EXERCISES

Exercise A   Reduce each fraction to its lowest terms.

1. $\dfrac{5}{15}$ 2. $\dfrac{6}{20}$ 3. $\dfrac{24}{44}$ 4. $\dfrac{36}{48}$ 5. $\dfrac{36}{84}$

Raise each fraction to a higher equivalent of each given denominator.

6. $\dfrac{4}{5} = \dfrac{}{25}$ 7. $\dfrac{5}{6} = \dfrac{}{30}$ 8. $\dfrac{11}{12} = \dfrac{}{48}$ 9. $\dfrac{7}{15} = \dfrac{}{60}$ 10. $\dfrac{11}{24} = \dfrac{}{72}$

Change these improper fractions to mixed numbers.

11. $\dfrac{19}{5}$ 12. $\dfrac{43}{12}$ 13. $\dfrac{27}{8}$ 14. $\dfrac{32}{18}$

15. $\dfrac{17}{2}$ 16. $\dfrac{45}{8}$ 17. $\dfrac{14}{13}$ 18. $\dfrac{16}{12}$

Change these mixed numbers to improper fractions.

19. $4\dfrac{3}{7}$ 20. $12\dfrac{1}{3}$ 21. $15\dfrac{2}{3}$ 22. $6\dfrac{3}{4}$

23. $2\dfrac{5}{7}$ 24. $7\dfrac{5}{6}$ 25. $12\dfrac{2}{3}$

Word Problems
Exercise B   Solve the following problems:

26. If the numerator and denominator of a fraction are the same, to what whole number is the fraction equal?

27. How would you write "15 divided by 20" as a fraction?

28. If five boys have to share $3, how much money will each one get?

29. A TV set selling for $75 was reduced by $\dfrac{1}{3}$. Write the amount of reduction as a fraction of the selling price.

30. Three partners had $30,000 to share equally.
    (a) What fraction of the money did each one get?
    (b) How many dollars did each one receive?

31. A sport jacket that cost the retailer $25 is marked up $10. Write as a fraction the markup compared to the cost.

32. In a shipment of glassware, 3 sets were damaged in shipping and 19 sets arrived undamaged.
    (a) What fraction of the shipment arrived damaged?
    (b) What fraction of the shipment arrived undamaged?

Reduce each answer to lowest terms.

33. Mary bought nine oranges. What fraction of a dozen did she buy?

34. A retailer bought 144 sport shirts and sold 48. What fraction of the 144 shirts was sold?

Unit 6: Addition and Subtraction of Fractions and Mixed Numbers

ADDITION OF FRACTIONS

Fractions can be added only if their denominators are the same.

  $\dfrac{5}{12} + \dfrac{7}{12} + \dfrac{3}{12} + \dfrac{8}{12}$

SOLUTION

The denominators are the same. The easiest way to solve this is to put the denominators together and add horizontally:

$$\frac{5 + 7 + 3 + 8}{12} = \frac{23}{12}$$

The result is an improper fraction, however. Change it to a mixed number.

$$\frac{23}{12} = 1\frac{11}{12}$$

  $\dfrac{2}{5} + \dfrac{1}{3} + \dfrac{11}{15}$

SOLUTION

The denominators are not the same, so we need to get a common denominator. Now, 5×3 is 15, so getting our common denominator is easy. It is 15. In the first fraction, we multiply the denominator (5) by 3 to get 15. Then we must multiply the numerator (2) by 3 to get 6. For the second fraction, we multiply the denominator (3) by 5 to get 15. We multiply the numerator (1) by 5 to get 5. We do not have to touch the third fraction. So we have:

$$\frac{6 + 5 + 11}{15} = \frac{22}{15}$$

Change it to a mixed number.

$$\frac{22}{15} = 1\frac{7}{15}$$

 EXAMPLE 3 $\dfrac{3}{4} + \dfrac{2}{3} + \dfrac{4}{5}$

SOLUTION

This is a difficult one, because all the denominators are different and most are prime. That is, they have no common factors. We will have to multiply all the denominators together to get the common denominator, $4 \times 3 \times 5 = 60$. So the common denominator is 60. The numerator for the first fraction must be multiplied by 15, because that is what the denominator (4) was multiplied by to get 60, $3 \times 15 = 45$. For the second fraction, the denominator (3) was multiplied by 20 to get 60. So the numerator (2) must be multiplied by 20, $20 \times 2 = 40$. For the third fraction, the denominator (5) was multiplied by 12 to get 60. So the numerator (4) must be multiplied by 12, $12 \times 4 = 48$. Now we are ready to add.

$$\frac{45 + 40 + 48}{60} = \frac{133}{60}$$

Change it to a mixed number.

$$\frac{133}{60} = 2\frac{13}{60}$$

 EXAMPLE 4 $2\dfrac{2}{5} + 4\dfrac{5}{6} + 7\dfrac{1}{3}$

SOLUTION

The best way to add mixed numbers is to add the whole parts first and then add the fractions. Change the result to a mixed number, and finally combine the whole part with the fractional part.

$$2 + 4 + 7 = 13.$$

The fractional parts are:

$$\frac{2}{5} + \frac{5}{6} + \frac{1}{3}$$

The common denominator is 30, because that is the lowest number common to all the denominators. For the first fraction, the denominator (5) was multiplied by 6 to get 30. So the numerator (2) must be multiplied by 6, $2 \times 6 = 12$. For the second fraction, the denominator (6) was multiplied by 5 to get 30. So the numerator (5) must be multiplied by 5, $5 \times 5 = 25$. For the third fraction, the denominator (3) was multiplied by 10 to get 30. So the numerator (1) must be multiplied by 10, $1 \times 10 = 10$. Now we are ready to add.

$$\frac{12 + 25 + 10}{30} = \frac{47}{30}$$

Change it to a mixed number.

$$\frac{47}{30} = 1\frac{17}{30}$$

Add this to the whole-number part:

$$13 + 1\frac{17}{30} = 14\frac{17}{30}$$

SUBTRACTION OF FRACTIONS

Subtraction is just the same as addition: we need a common denominator to subtract fractions.

 EXAMPLE 5 $\dfrac{7}{9} - \dfrac{6}{11}$

SOLUTION

The denominators are not the same. This equation gives us a chance to see that, often, the two denominators must be multiplied together to add or subtract. The common denominator is 99 (multiplying the two denominators). For the first fraction, since we multiplied 9 by 11 to get 99, we multiply 7 by 11 to get 77. For the second fraction, since we multiplied 11 by 9 to get 99, we multiply 6 by 9 to get 54. Now we can do the subtraction:

$$\frac{77}{99} - \frac{54}{99} = \frac{23}{99}$$

This fraction cannot be reduced. There is no number that can be divided into both 23 and 99. In fact, 23 is a prime number.

SUBTRACTION OF MIXED NUMBERS

Subtract the whole numbers and then the fractions.

 EXAMPLE 6 $34\dfrac{5}{8} - 22\dfrac{5}{6}$

SOLUTION

Subtract the whole numbers first:

$$34 - 22 = 12$$

Then subtract the fractional parts:

$$\frac{5}{8} - \frac{5}{6}$$

The common denominator is 24 (the smallest number that divides both 8 and 6). For the first fraction, since we multiplied 8 by 3 to get 24, we multiply 5 by 3 to get 15. For the second fraction, since we multiplied 6 by 4 to get 24, we multiply 5 by 4 to get 20. Now we can do the subtraction:

$$\frac{15}{24} - \frac{20}{24}$$

Look out. The second fraction is larger than the first. We need to borrow one from the whole numbers, so 12 becomes 11. We then add the fraction 24/24 to 15/24 and get:

$$\frac{39}{24} - \frac{20}{24} = \frac{19}{24}$$

So we see that

$$34\frac{5}{8} - 22\frac{5}{6} = 11\frac{19}{25}$$

EXERCISES

Exercise A   Solve the following problems:

1.
$$\begin{array}{r} \frac{3}{5} \\ \frac{2}{3} \\ +\frac{9}{15} \\ \hline \end{array}$$

2.
$$\begin{array}{r} \frac{1}{4} \\ \frac{5}{8} \\ +\frac{13}{24} \\ \hline \end{array}$$

3.
$$\begin{array}{r} \frac{2}{8} \\ \frac{2}{3} \\ +\frac{5}{6} \\ \hline \end{array}$$

4.
$$\begin{array}{r} 12\frac{3}{8} \\ 5\frac{2}{3} \\ +19\frac{6}{7} \\ \hline \end{array}$$

5.
$$\begin{array}{r} 35\frac{2}{3} \\ 15\frac{13}{15} \\ +18\frac{3}{4} \\ \hline \end{array}$$

6.
$$\begin{array}{r} 23\frac{1}{2} \\ 19\frac{7}{8} \\ +21\frac{4}{5} \\ \hline \end{array}$$

7.
$$\begin{array}{r} 15\frac{2}{3} \\ 18\frac{4}{5} \\ +24\frac{5}{6} \\ \hline \end{array}$$

8.
$$\begin{array}{r} 14\frac{2}{3} \\ 20\frac{4}{5} \\ +17\frac{1}{2} \\ \hline \end{array}$$

9.
$$\begin{array}{r} \frac{4}{5} \\ -\frac{3}{4} \\ \hline \end{array}$$

10.
$$\begin{array}{r} \frac{3}{5} \\ -\frac{5}{12} \\ \hline \end{array}$$

11. $\dfrac{3}{4}$
$-\dfrac{2}{3}$

12. 19
$-3\dfrac{7}{8}$

13. $37\dfrac{1}{5}$
$-20\dfrac{5}{6}$

14. $23\dfrac{3}{8}$
$-18\dfrac{2}{3}$

15. $15\dfrac{2}{3}$
$-7\dfrac{5}{12}$

Exercise B Complete the following forms:

16.

		Hours Worked						Total
Card No.	Name of Employee	M	T	W	T	F	S	Total Hours Worked
40	S. Alvarez	$8\frac{1}{2}$	$7\frac{3}{4}$	9	$8\frac{1}{4}$	$8\frac{3}{4}$	4	____
41	W. Baines	9	$8\frac{1}{4}$	$9\frac{1}{2}$	$7\frac{3}{4}$	$8\frac{1}{2}$	3	____
42	P. Belmore	$8\frac{3}{4}$	$9\frac{1}{2}$	$8\frac{1}{4}$	$8\frac{3}{4}$	$9\frac{3}{4}$	5	____
43	L. Caldwell	$9\frac{1}{2}$	$8\frac{3}{4}$	$8\frac{1}{4}$	$8\frac{3}{4}$	$9\frac{1}{2}$	4	____
44	N. Carter	$8\frac{1}{2}$	$9\frac{3}{4}$	$8\frac{1}{4}$	$8\frac{3}{4}$	$9\frac{1}{2}$	4	____
45	I. Cortez	$9\frac{1}{2}$	$8\frac{3}{4}$	$8\frac{1}{2}$	$9\frac{3}{4}$	$8\frac{1}{4}$	5	____
46	V. Elton	$8\frac{1}{2}$	$9\frac{3}{4}$	$8\frac{1}{4}$	$8\frac{3}{4}$	$9\frac{1}{2}$	4	____

Week Ending June 14, 20–

17.

		Hours Worked						Total
Card No.	Name of Employee	M	T	W	T	F	S	Total Hours Worked
40	S. Alvarez	$8\frac{1}{2}$	$9\frac{3}{4}$	$9\frac{1}{2}$	$8\frac{1}{4}$	$8\frac{3}{4}$	4	____
41	W. Baines	$9\frac{3}{4}$	$8\frac{1}{2}$	$8\frac{1}{4}$	$9\frac{3}{4}$	$8\frac{1}{4}$	5	____
42	P. Belmore	$8\frac{3}{4}$	$9\frac{1}{2}$	$8\frac{1}{4}$	$9\frac{3}{4}$	$8\frac{1}{2}$	4	____
43	L. Caldwell	$9\frac{3}{4}$	$8\frac{3}{4}$	$8\frac{1}{2}$	$9\frac{1}{4}$	$8\frac{3}{4}$	4	____
44	N. Carter	$9\frac{1}{2}$	$9\frac{3}{4}$	$8\frac{1}{2}$	$9\frac{1}{4}$	$8\frac{3}{4}$	5	____
45	I. Cortez	$8\frac{3}{4}$	$9\frac{1}{2}$	$8\frac{1}{2}$	$9\frac{1}{4}$	$8\frac{3}{4}$	4	____
46	V. Elton	$8\frac{1}{2}$	$9\frac{3}{4}$	$8\frac{1}{4}$	$9\frac{1}{2}$	$9\frac{3}{4}$	4	____

Week Ending June 28, 20–

Word Problems

Exercise C   Solve the following problems:

18. A salesclerk sold the following pieces of material: $4\frac{1}{2}$ yards, $3\frac{1}{4}$ yards, $6\frac{2}{3}$ yards, and $5\frac{5}{6}$ yards. How many yards did the salesclerk sell?

19. A manufacturer needs the following pieces of material to make a suit: $1\frac{1}{2}$ yards for the jacket, $1\frac{2}{3}$ yards for the pants, and $\frac{4}{6}$ yard for the vest. Find the total number of yards needed to make the suit.

20. A mixture contains three ingredients, present in the following amounts: $5\frac{1}{2}$ pounds, $3\frac{1}{3}$ pounds, and $8\frac{7}{16}$ pounds. What is the total weight of the mixture?

21. Mr. Brandt bought $\frac{5}{6}$ of a ton of coal and used $\frac{2}{3}$ ton. How much coal was left?

22. A carpenter had a piece of board measuring $24\frac{1}{2}$ feet. If he cut off a piece measuring $6\frac{2}{3}$ feet, how many feet of board were left?

Unit 7: Multiplication and Division of Fractions and Mixed Numbers

MULTIPLICATION OF FRACTIONS

Multiplication of fractions is rather easy. We simply multiply the numerators and the denominators:

$$\frac{7}{9} \times \frac{6}{7} = \frac{42}{63}$$

SOLUTION

We can reduce this fraction. If we divide both the numerator and the denominator by 3 and then by 7, the fraction reduces to:

$$\frac{2}{3}$$

This is a good time to show that we could have reduced before we multiplied:

$$\frac{{}^1\cancel{7}}{{}_3\cancel{9}} \times \frac{\cancel{6}^2}{\cancel{7}_1} = \frac{2}{3}$$

Reducing before multiplying makes the problem much simpler. We can reduce up and down, and also diagonally across. Always reduce, as it makes the resulting multiplication easier. We will see this in the next example.

 EXAMPLE 2

$$\frac{3}{7} \times \frac{5}{6} \times \frac{4}{10}$$

SOLUTION

By reducing all the fractions we can, we have:

$$\frac{{}^1\cancel{3}}{7} \times \frac{{}^1\cancel{5}}{{}^1\cancel{6}} \times \frac{{}^1\cancel{4}}{{}^1\cancel{10}} = \frac{1}{7}$$

You can see that reducing the fractions makes an enormous difference in difficulty of the calculations.

 EXAMPLE 3

$$\frac{5}{32} \times \frac{7}{24} \times \frac{6}{25}$$

SOLUTION

This is more challenging because we cannot reduce as much. Here are the reductions:

$$\frac{{}^1\cancel{5}}{32} \times \frac{7}{{}_4\cancel{24}} \times \frac{{}^1\cancel{6}}{{}_5\cancel{25}}$$

The fractions, thus reduced, become:

$$\frac{1}{32} \times \frac{7}{4} \times \frac{1}{5} = \frac{7}{640}$$

You might think that the product is a rather small fraction and something that you would not see. However, fractions like this come up all the time in business, particularly in leases.

$$2\frac{5}{9} \times 13\frac{7}{11}$$

SOLUTION

The first thing to do when multiplying mixed numbers is to convert each mixed number to an improper fraction. In the first fraction, multiply the whole number (2) by the denominator (9): $2 \times 9 = 18$ and add to 5: $18 + 5 = 23$. In the second fraction, we have $13 \times 11 = 143$ and $143 + 7 = 150$. So the revised fractions are:

$$\frac{23}{9} \times \frac{150}{11} = \frac{3,450}{99}$$

This fraction cannot be reduced. However, it can be changed to a mixed number. Divide 3,450 by 99 and get 34.84848484. The whole-number part is 34. Then multiply 0.848484 by 99 and get 84. (Actually, on a calculator, you get 83.9999999, but we round this to 84.) So the final answer is:

$$34\frac{84}{99}$$

We can reduce the fraction by dividing both numerator and denominator by 3:

$$34\frac{28}{33}$$

$$5\frac{7}{8} \times 3\frac{4}{5} \times 17\frac{3}{8}$$

SOLUTION

First, we convert all the mixed numbers to improper fractions. In the first fraction, $5 \times 8 = 40$ and $40 + 7 = 47$. In the second, $3 \times 5 = 15$ and $15 + 4 = 19$. In the third, $17 \times 8 = 136$ and $136 + 3 = 139$. So the converted fractions and multiplication look like this:

$$\frac{47}{8} \times \frac{19}{5} \times \frac{139}{8} = \frac{124,127}{320}$$

We must convert this improper fraction to a mixed number. Dividing 124,127 by 320 results in 387.89687. Then multiplying 0.89687 by 320 results in 287 (after we round up). So the final mixed number is:

$$387\frac{287}{320}$$

DIVISION OF FRACTIONS

Dividing fractions is very similar to multiplying fractions. There is one twist, though. The second fraction must be inverted and multiplied by the first fraction. *Invert* means to take the reciprocal, that is, make the numerator the denominator and make the denominator the numerator. For example:

$$\frac{7}{8} \quad \text{inverted becomes} \quad \frac{8}{7}$$

The procedure, then, is to invert the second fraction and then multiply. Remember, though, that the *second* fraction needs to be inverted, never the first.

 EXAMPLE 6 $\frac{10}{13} \div \frac{35}{39}$ becomes $\frac{10}{13} \times \frac{39}{35}$

Now we can reduce:

$$\frac{\overset{2}{\cancel{10}}}{\underset{1}{\cancel{13}}} \times \frac{\overset{3}{\cancel{39}}}{\underset{7}{\cancel{35}}}$$

New multiply to get the reduced answer:

$$\frac{2}{1} \times \frac{3}{7} = \frac{6}{7}$$

Again, note how easy the problem becomes if it can be reduced before actually multiplying.

DIVISION OF MIXED NUMBERS

When dividing mixed numbers, you first have to convert to an improper fraction.

 EXAMPLE 7 $3\frac{4}{6} \div 6\frac{4}{9}$

SOLUTION

First, convert the mixed numbers to improper fractions. For the first fraction, $3 \times 6 = 18 + 4 = 22$. For the second fraction, $6 \times 9 = 54 + 4 = 58$. So the division becomes:

$$\frac{22}{6} \div \frac{58}{9}$$

After inverting the second fraction, insert a multiplication sign:

$$\frac{22}{6} \times \frac{9}{58}$$

Reduce top and bottom to get:

$$\frac{\overset{11}{\cancel{22}}}{\underset{2}{\cancel{6}}} \times \frac{\overset{3}{\cancel{9}}}{\underset{29}{\cancel{58}}}$$

The reduced answer is:

$$\frac{11}{2} \times \frac{3}{29} = \frac{33}{58}$$

This answer cannot be reduced any further.

CHECKING FRACTION OPERATIONS

Most calculators will not do fraction operations, at least not business calculators. Most calculators that do fraction operations are specifically made for elementary school.

EXERCISES

Exercise A Solve the following problems:

1. $\frac{7}{8} \times \frac{14}{21}$

2. $\frac{7}{16} \times \frac{2}{3} \times \frac{4}{9}$

3. $\frac{4}{5} \times \frac{7}{8} \times \frac{2}{3}$

4. $6\frac{1}{2} \times 3\frac{1}{3}$

5. $12 \times 2\frac{1}{5} \times \frac{1}{2}$

6. $\frac{5}{8} \div \frac{3}{4}$

7. $\frac{2}{3} \div \frac{9}{12}$

8. $\frac{3}{5} \div \frac{5}{20}$

9. $15\frac{2}{3} \div 8\frac{5}{9}$

10. $8\frac{1}{2} \div 4$

11. $4\frac{3}{4} \div 5\frac{1}{2}$

12. $8 \div \frac{3}{4}$

13. $14\frac{1}{2} \div 3\frac{5}{8}$

Word Problems

Exercise B   Solve the following problems:

14. How many $\frac{1}{5}$'s of a yard are there in $\frac{3}{4}$ of a yard?

15. A plane flies 980 miles in $1\frac{2}{3}$ hours. How many miles does the plane average in 1 hour?

16. A bolt of material contains 45 yards of cotton. If it takes $2\frac{1}{5}$ yards to make a dress, how many dresses can be made from the bolt of cotton?

17. A TV selling for $636.45 was reduced by $\frac{1}{3}$. Find the amount of reduction.

18. John earns $250 a week. He spends $\frac{1}{3}$ on rent and utilities, spends $\frac{1}{5}$ on food, and saves the rest. How much money does John spend on each of these items: (a) rent and utilities, (b) food, (c) savings?

Unit 8: Review of Chapter 1

TERMS:	• Addends
	• Sum or Total
KEY PHRASES:	• "What is the total?"
	• "Find the (total) cost of . . . ."
HINTS:	• Line up the decimal points in vertical addition.
	• Check your totals.

Find the totals on the following forms:

1.

Please endorse all checks and list below singly

DATE	,20	DOLLARS	CENTS
	BILLS	428	00
	COINS	23	86
	CHECKS	73	68
		110	52
		93	87
		47	58
		242	34
TOTAL			

2.

Please endorse all checks and list below singly

DATE	,20	DOLLARS	CENTS
	BILLS	928	00
	COINS	123	86
	CHECKS	93	69
		128	82
		95	04
		69	58
		193	94
TOTAL			

3.

Sales for June				
Department	Cash Sales		Charge Sales	
A	$14,728	63	$15,628	47
B	8,649	47	9,473	64
C	29,382	69	32,592	63
D	17,461	63	19,637	46
E	31,728	41	34,528	47
F	20,649	47	21,473	64
G	15,349	72	16,593	75
Totals				

4.

Salesperson	Sales		Commissions	
A	$23,575	85	$1,650	31
B	21,728	60	1,521	00
C	19,535	25	1,367	47
D	27,382	49	1,916	77
E	24,628	35	1,723	98
Totals				

5.

Petty Cash Expenditures		
Date	Amount	
9/13	$15	83
9/16	19	65
9/17	25	27
9/18	11	45
9/19	27	78
9/20	39	57
9/23	23	35
9/24	16	42
9/25	27	35
Total		

Solve the following problem:

6. James Washington earned the following commissions: $1,463.75; $1,634.80; $1,678.65; and $1,273.32. What was the sum of all his commissions?

TERMS:
- Minuend
- Subtrahend
- Difference

KEY PHRASES:
- "What is the difference?"
- "How much greater is . . . ?"
- "How much less is . . . ?"
- "What is the net . . . ?"
- "By how much does something exceed . . . ?"

HINTS:
- Line up the decimal points in vertical subtraction.
- Check all subtraction.

Complete the following forms:

7.

Item	Original Price		Sales Price		Amount of Reduction	
A	$305	60	$248	75	_____	___
B	310	40	142	85	_____	___
C	508	00	329	95	_____	___
D	400	00	325	26	_____	___
E	705	05	584	88	_____	___
F	740	00	538	75	_____	___
G	825	90	634	95	_____	___
Totals		−		=		**Grand Total**

8.

Item	List Price		Discount		Net Price	
A	$248	95	$99	95	_____	___
B	163	00	19	99	_____	___
C	55	50	9	50	_____	___
D	235	75	29	75	_____	___
E	385	00	95	00	_____	___
F	82	48	19	50	_____	___
G	136	00	39	95	_____	___
Totals		−		=		**Grand Total**

9.

List Price	$468	92
Trade Discount	194	68
Net Price		

10.

List Price	$12,312	97
Trade Discount	5,460	98
Net Price		

11.

Invoice No.	Amount		Cash Discount		Net Amount	
4137	$ 1,284	65	$ 28	74	_____	___
4332	3,524	32	75	28	_____	___
4253	12,863	75	253	68	_____	___
4315	13,615	92	273	95	_____	___
4528	15,225	63	342	75	_____	___
Totals		−		=	**Grand Total**	

12.

Card No.	No. of Exemptions	Total Wages		Deductions									Total Deductions		Net Pay	
				Soc. Sec.		Fed. With. Tax		State Tax		City Tax						
101	2	$465	85	$32	61	$116	46	$28	53	$12	65		___	___	___	___
102	1	432	70	30	28	109	32	26	72	11	28		___	___	___	___
103	3	478	48	34	63	123	16	29	43	12	84		___	___	___	___
104	1	398	42	29	72	103	15	24	82	9	80		___	___	___	___
105	2	446	36	31	16	113	70	27	16	10	80		___	___	___	___
106	4	480	70	34	32	102	62	26	42	11	73		___	___	___	___

Solve the following problems:

13. Samantha earns $19,500 a year. Her federal, state, and city income taxes amount to $4,763.85. What is her net salary for the year?

14. Sandy earns $1,060.05 a month. Her monthly expenses are as follows: rent, $345; food, $185.75; gas and electric, $85.95; car expenses, $79.60; other expenses, $135.45. How much of her salary is left after expenses?

TERMS:
- Multiplicand
- Multiplier
- Product

KEY PHRASES:
- "What is the total?"
- "Find the cost of . . . ."

HINTS:
- Set up your problem neatly on scratch paper.
- Align all columns of digits carefully.
- Check your work.
- Use shortcuts when multiplying with powers of 10.
- Round numbers when necessary.
- Use multiplication as a shortcut for addition whenever the addends are the same.

Complete the following forms:

15.

Quantity	Description	Unit Price		Extension	
18.25 tons	Rice	$123	62	_____	___
24.262 tons	Corn	93	85	_____	___
15.032 tons	Lentil	131	75	_____	___
23.7 tons	Soya Bean	87	43	_____	___
13.623 tons	Sesame	78	65	_____	___
		Total			

16.

Quantity	Description	Unit Price		Extension	
500 boxes	#10 Paper Clips	$	20	_____	___
125 boxes	#063 Typ. Ribbons, Blue/Red	20	00	_____	___
2,000 bottles	Correction Fluid		70	_____	___
300 boxes	#6 Envelopes	1	20	_____	___
100 boxes	Desk Pads	14	60	_____	___
		Total			

Word Problems

 Samantha bought a new car and made a down payment of $1,800. If she pays the balance in 60 payments of $200 each, what is the total cost of the car?

SOLUTION

Solve the problem using the shortcut method of multiplication by a power of 10.

GIVEN FACTS: 1. Samantha made a down payment of $1,800.
2. She will make 60 payments of $200 each.

FIND: The total cost of the car.
Total cost is $1,800, plus 60 payments of $200 each.

Total cost = $1,800 + (60 × $200)
= 1,800 + (6 × 2 increased by 3 zeros)
= 1,800 + 12,000
= 13,800
Total cost = $13,800

CHECK: 1800 ⊞ ⟦60 ⊠ 200⟧ ⊟ 13800✓
ANSWER: Total cost of Samantha's car is $13,800.

Solve the following problems:

17. Find the cost of 2,000 springs priced at $0.055 each.

18. How much will 100 filters cost at $12.50 each?

19. The Taylor Real Estate Company purchased eight electric typewriters at $972.50 each. What was the total cost of the typewriters?

20. Aldo bought a new car, to be paid for in 36 monthly installments. Each monthly payment is $275.65. If his down payment was $500, what is the total cost of the car?

21. Acme Motorcycles sells a motorcycle for $99 down and 36 monthly payments of $49.95. What is the total cost of the motorcycle?

22. Charles worked 54 hours last week. His hourly rate of pay is $4.75 for regular hours and $7.125 for all hours worked beyond 40 hours. What is his total pay for the week?

23. Alice is paid a 7% commission on sales. How much is her commission on a sale of $638.75?

TERMS:
- Dividend
- Divisor
- Quotient
- Remainder

KEY PHRASES:
- "What is the unit cost?"
- "How much is each item?"
- "What is the . . . per . . . ?"

HINTS:
- Read the problem carefully.
- Set up your work neatly, and leave enough space.
- Check your division by multiplication.
- Remember to use power of 10 shortcuts.
- Round your answer if necessary.

Find the quotient in each of the following problems:

24. $48\overline{)7,4976}$ 25. $478\overline{)2,1988}$ 26. $38\overline{)81.70}$ 27. $0.015\overline{)6,3195}$

Find the quotient in each of the following problems, and round to the nearest *cent*:

28. $24\overline{)\$590.52}$ 29. $215\overline{)\$4,853.42}$ 30. $321\overline{)\$43,460.05}$

Find the quotient in each of the following problems. (Be alert for zeros in the quotient.)

31. $64\overline{)32,320}$ 32. $56\overline{)224,168}$ 33. $63\overline{)1,260.252}$ 34. $24\overline{)15,128}$

Find the quotient in each of the following problems. (In problems involving dollars and cents, round to the nearest *cent*.)

35. $14.75 ÷ 100

36. 3,000 ÷ 500

37. 600 ÷ 30

38. 2,467 ÷ 1,000

39. 3,000 ÷ 1,500

Complete the following forms, rounding each number to the nearest *cent*.

40.

Installment Price		Down Payment		Number of Payments	Amount of Each Payment	
$ 865	95	$ 75	00	12	_____	___
1,535	45	150	00	18	_____	___
2,792	48	200	00	24	_____	___
5,478	90	550	00	36	_____	___
10,563	75	650	00	48	_____	___

41.

Quantity	Description	Price	Amount	
675 lb.	Sugar	$4.15 per cwt.	_____	___
1,275 lb.	Coffee	$6.45 per cwt.	_____	___
8,460 lb.	Wheat	$267.25 per ton	_____	___
12,250 lb.	Peanuts	$325.95 per ton	_____	___

Solve the following problems:

42. The First Federal Savings Bank bought 15 electronic calculators for $5,189.25. What was the cost of each calculator?

43. The Hartford Insurance Company ordered 1,685 calendars priced at $5.35 per 100, and 8,490 imprinted pens priced at $15.25 per 1,000. What was the total cost of the order?

44. Mr. Gunfield bought 200 shares of stock at $63.75 each share. Two years later he sold the stock at a price of $86.25 each share. How much profit did he make?

TERMS:
- Common fraction
- Complex fraction
- Decimal fraction
- Denominator
- Equivalent fraction
- Improper fraction
- Mixed number
- Numerator

HINTS:
- Reduce all answers.
- Check your work.
- Add and subtract with common denominators *only*.
- Use decimal and common fraction equivalents.
- Use the $\dfrac{IS}{OF}$ fraction and the basic function equation to solve fraction problems.

Reduce each fraction to lowest terms.

45. $\dfrac{3}{24}$ 46. $\dfrac{8}{18}$ 47. $\dfrac{30}{300}$

Raise each fraction to the higher equivalent of each denominator.

48. $\dfrac{3}{4} = \dfrac{}{48}$ 49. $\dfrac{12}{15} = \dfrac{}{45}$

Add the following fractions and mixed numbers:

50. $\dfrac{3}{4}$
$\dfrac{4}{5}$
$+\dfrac{7}{10}$

51. $\dfrac{7}{15}$
$\dfrac{5}{6}$
$+\dfrac{4}{9}$

52. $19\dfrac{5}{8}$
$32\dfrac{3}{5}$
$+28\dfrac{13}{20}$

Subtract the following fractions and mixed numbers:

53. $\dfrac{2}{3}$
$-\dfrac{3}{5}$

54. $\dfrac{5}{6}$
$-\dfrac{4}{16}$

55. 20
$-14\dfrac{7}{8}$

Multiply the following fractions and mixed numbers:

56. $\dfrac{4}{5} \times \dfrac{15}{16}$

57. $5\dfrac{1}{2} \times 6\dfrac{2}{3} \times 4\dfrac{4}{5}$

58.
$$
\begin{array}{r}
236 \\
\times 35\dfrac{1}{4} \\
\hline
\end{array}
$$

Divide the following fractions and mixed numbers:

59. $\dfrac{7}{8} \div \dfrac{15}{16}$

60. $2\dfrac{1}{5} \div 4\dfrac{3}{10}$

61. $4\dfrac{3}{5} \div 5\dfrac{7}{15}$

Decimals, Percents, Ratios, and Proportions

Unit 1: Decimals in Business

In business, numbers are often shown as decimals, fractions, and percents.

½ Fraction
0.5 Decimal
50% Percent

More than a whole can also be shown. Suppose we have one and one-quarter containers of plasma TVs. The number of containers looks like this:

1¼ Fraction
1.25 Decimal
125% Percent

A decimal is a base 10 way of showing a part of a whole. Some examples:

0.9 Nine-tenths
0.43 Forty-three hundredths
0.059 Fifty-nine thousandths

The fourth place to the right the decimal place is ten-thousandth, and so forth.

CHANGING DECIMALS TO FRACTIONS

To change a decimal to a fraction, simply put the numerator to the right of the decimal point in the correct position. The denominator tells you which position is correct. Let's use the examples above:

$$0.9 = \frac{9}{10} \qquad 0.43 = \frac{43}{100} \qquad 0.059 = \frac{59}{1,000}$$

Sometimes we need to reduce the fraction. For instance, 0.5 = 5/10, which reduces to 1/2. Another example is 0.25 = 25/100, which reduces to 1/4.

CHANGING FRACTIONS TO DECIMALS

To change a fraction to a decimal, simply divide the numerator by the denominator.

 Change 1/4 to a decimal.

$$
\begin{array}{r}
0.25 \\
4{\overline{\smash{\big)}\,1.00}} \\
\underline{.8} \\
20 \\
\underline{20} \\
0
\end{array}
$$

Note that 4 goes into 1.00 evenly, so the decimal equivalent of 1/4 is 0.25. However, what happens when the numerator and denominator do not divide evenly? In that case, we need to estimate the decimal. For instance, what is the decimal equivalent of 1/3? Doing the division results in:

$$
\begin{array}{r}
0.333 \\
3{\overline{\smash{\big)}\,1.00}} \\
\underline{.9} \\
10 \\
\underline{9} \\
10 \\
\underline{9} \\
10
\end{array}
$$

Note that 3 will always be the quotient, because 1 is always the remainder after subtracting of 9 from 10. When this happens, we simply put 0.333, or we could put down:

$$0.\overline{33}$$

In other words, we could write the 3 with a line over it. This means that the 3 repeats again and again.

 Change $\dfrac{2}{3}$ to a decimal.

$$
\begin{array}{r}
0.666 \\
3{\overline{\smash{\big)}\,2.00}} \\
\underline{.18} \\
20 \\
\underline{18} \\
20 \\
\underline{18} \\
2
\end{array}
$$

What can be done with 0.66 (repeating forever)? One way to write it is 0.67, rounding off. The other way is to write it the way we did with 1/3:

$$0.\overline{66}$$

When converting from a fraction to a decimal, rounding off a decimal to the nearest hundredth is adequate for most business situations. However, it does mean you have to divide to the thousandths and then round off.

The following tables show some of the most frequently used fractions and their decimal equivalents.

TABLE 2.1
COMMON FRACTION-DECIMAL EQUIVALENTS

Fraction	Decimal
1/4	0.25
1/3	0.33333
1/2	0.5
2/3	0.66666
3/4	0.75

TABLE 2.2
COMMON EQUIVALENTS OF EIGHTHS

Fraction	Decimal
1/8	0.125
$\frac{2}{8} = \frac{1}{4}$	0.25
3/8	0.375
$\frac{4}{8} = \frac{1}{2}$	0.5
5/8	0.625
$\frac{6}{8} = \frac{3}{4}$	0.75
7/8	0.875
1	1.0

The eighths are useful because three of them can be reduced: $\frac{2}{8} = \frac{1}{4}, \frac{4}{8} = \frac{1}{2}, \frac{6}{8} = \frac{3}{4}$. Eighths are also useful because many applications require them. Consider memorizing their fraction-decimal equivalents.

EXERCISES

Exercise A Change the following fractions to decimal fractions to the nearest *tenth*:

1. $\dfrac{5}{7}$ 2. $\dfrac{8}{15}$ 3. $\dfrac{12}{17}$

Change the following common fractions to decimal fractions to the nearest *hundredth*:

4. $\dfrac{7}{8}$ 5. $\dfrac{19}{45}$ 6. $\dfrac{34}{47}$

Change the following common fractions to decimal fractions to the nearest *thousandth*:

7. $\dfrac{18}{34}$ 8. $\dfrac{15}{54}$ 9. $\dfrac{31}{43}$

Word Problems

Exercise B Solve the following problems:

10. A computer gear measures $\dfrac{21}{32}$ inch. How many thousandths of an inch does the gear measure?

11. A part is produced with a tolerance of "plus or minus $\dfrac{3}{64}$ of an inch." How many thousandths of an inch is the tolerance?

12. How many tenths of an hour is $\dfrac{5}{6}$ hour?

13. Pam bought 2 feet of fabric. How *many thousandths* of a yard did she buy? (One yard equals 3 feet.)

14. Twelve seconds is what fraction of a minute in decimals?

Unit 2: Percents in Business

Percents are used in every aspect of business. As mentioned before, though, percents are another way of showing a fraction, or decimal. *Percent* literally means "per hundred." We use percents in all aspects of business. Some examples include:

- A 6% tax on $10.00 would be 60 cents ($0.60).
- A salesperson getting a 15% commission on a car sale will make $15 on each $100. So if the car costs $10,000.00, the salesperson will make $1500 as a commission.
- If there is a 30% reduction sale on lawn mowers, for each $100, the price is reduced by $30. So if a lawn mower originally sold for $150, the reduction would be $45 and the sale price would be $105.
- If a theater holds 1500 people and it is 70% full, then for every 100 seats, 70 people are in the theater. So the theater would be holding 1050 people.

CHANGING DECIMALS TO PERCENTS

To change a decimal to a percent, simply move the decimal place two places to the right and put a percent sign on the resulting number.

Decimal: 0.12	Percent: 12%
Decimal: 0.425	Percent: 42.5%
Decimal: 0.006	Percent: 0.6%

Note that in the last two decimal to percent conversions, we have tenths of a percent. This happens all the time in business.

CHANGING PERCENTS TO DECIMALS

To change a percent to a decimal, simply move the decimal place two places to the left and drop the percent sign:

Percent: 68%	Decimal: 0.68
Percent: 12.5%	Decimal: 0.125
Percent: 125%	Decimal: 1.25

Note that sometimes you have more than 100%. This results in a decimal that is more than one. For instance, 125% equals 1.25 (or 1¼ as a fraction). Remember that fractions, decimals, and percents are all ways of looking at the same thing: parts of a whole.

IDENTIFYING THE PARTS OF A PERCENT EQUATION

In an equation dealing with percents, it is useful to know the parts of the equation.

 Jeanne bought $50.00 worth of electronics and paid 7% sales tax on it. How much did she pay in sales tax?

SOLUTION

Here is the basic percent equation (and the names of the parts):

$$\text{Base} \times \text{Rate} = \text{Percentage}$$
$$\$50 \times 0.07 = \$3.50$$

Note that it is usually easiest to convert the percent rate to a decimal. That way, the calculator can handle the resulting multiplication very quickly.

 Lydia, who sells houses, earns a 12% commission on every house she sells. If she sells a house for $155,000, how much commission will she earn?

SOLUTION

$$\text{Base (house price)} \times \text{Rate} = \text{Percentage (commission)}$$
$$\$155,000 \times 0.12 = \$18,600$$

 Donna calculates federal taxes for an accounting firm. She earns an 8% commission on every tax return she prepares. If she prepares a client's taxes and the firm's bill is $255, how much commission will she earn?

SOLUTION

$$\text{Base (bill)} \times \text{Rate} = \text{Percentage (commission)}$$
$$\$255 \times 0.08 = \$20.40$$

 Bob bought a bomber jacket. The price of the jacket was $279.99. If 5% sales tax was added, what was the total price Bob paid?

SOLUTION

$$\text{Base (price)} \times \text{Rate} = \text{Percentage (Tax amount)}$$
$$\$279.99 \times 0.05 = \$13.9995$$

Note that we get a decimal out to the ten-thousandths place. We round off to the hundredths place (the nearest penny) and get the tax amount of $14.00. Then we have:

$$\$279.99 + \$14.00 = \$293.99$$

$293.99 is the amount that Bob will need to pay for the jacket.

EXERCISES

Exercise A   Convert decimals to percents and percents to decimals:

1. 29%　　　2. 37.5%　　　3. 100%　　　4. 7.98%　　　5. 0.01%

6. 6.75%　　　7. 0.25　　　8. 0.06　　　9. 87.5　　　10. 2

11. 2.15　　　12. 0.20　　　13. $8\frac{1}{4}$%　　　14. $12\frac{1}{2}$%

Find the percentage in each of the following problems:

15. 35% of $728　　　16. 37.5% of $546　　　17. 7.5% of $2,400

18. 10% of $425.60　　　19. 8% of $173.75　　　20. 5% of $150.60

Exercise B   Complete the following form:

21.

PENNY LANE SHOP

February 3　20--

SOLD TO ___ *S. Tipsley* ___

ADDRESS ___ *2110 Park Place* ___

CLERK *S.T.*	DEP'T *012*	AM'T REC'D $ *1,265*

QUAN.	DESCRIPTION	AMOUNT	
1	*Recliner Chair*	$258	00
1	*Desk*	576	00
1	*Desk Chair*	87	00
1	*Bookcase*	246	00
	Subtotal		
	8% Sales tax		
	Total		

POSITIVELY NO EXCHANGES MADE UNLESS
THIS SLIP IS PRESENTED WITHIN 3 DAYS.

Word Problems

Any problem asking for a percentage must supply you with a rate and a base. Most often, such problems will ask for the sales tax or the discount amount.

"What is the tax on sales of $1,500 if the tax rate is 5%?"

"Find the net price if the list price is $15.95 and the discount rate is 25%."

"What is the cost overrun if the contract price was $1,100 and the overrun rate was 7.8%?"

Exercise C Solve the following problems:

22. The price of a stereo set was reduced by 40%. How many cents out of every dollar of the price represents the reduction?

23. If you save $33\frac{1}{3}$ cents on every dollar of the cost of a yard of carpeting, what is the percent of the savings?

24. A bank pays $7.50 in interest for every $100 of savings. What is the percent of interest?

25. Mr. Wells purchased a home for $48,990. If his down payment is 20%, find the amount of his mortgage.

26. Sally earns $1,175 a month. If she spends 20% on rent and 15% on food, how much of her monthly income remains?

27. A salesperson earns a commission of 15%. What is the commission on sales of $2,530?

AMOUNT OF INCREASE OR DECREASE IN A PERCENT EQUATION

In business it is important to be able to calculate the percent from the amounts, or the percent of a decrease.

TRANSLATING THE IS/OF PROBLEM INTO AN EQUATION

To solve a percent equation, it is important to translate the words into an equation.

A good memory device is to use the fraction $\frac{IS}{OF}$.

The way to translate this into an equation is to remember that the number related to the IS is the numerator of the equation. The number related to the OF is the denominator of the equation. The percent is always placed over 100 because it is "per hundred". The following examples show the three kinds of equations encountered.

20 is what percent of 30?

SOLUTION

In this problem, we are looking for the percent, so that is the unknown. 20 is in the numerator because it is related to the IS in the problem. 30 is in the denominator because it is related to the OF in the equation. So we set up the proportion:

$$\frac{20}{30} = \frac{x}{100}$$

Because we are looking for the percent, that is the unknown. The percent is over 100, which is the definition of percent, as was noted earlier. To solve this, we cross multiply:

$$30x = (20)(100) \Rightarrow x = 2{,}000/30 \Rightarrow x = 66.67$$

Since x is a percent, 20 is 66.67% of 30.

45% of what number is 23?

SOLUTION

45 is a percent, so it is put over 100. 45 is across the equal sign from the 23 because 23 is related to the IS. The unknown is across from the 100 because it is related to the OF:

$$\frac{45}{100} = \frac{23}{x}$$

By cross multiplying, we get $45x = 2{,}300 \Rightarrow x = 2{,}300/45 \Rightarrow x = 51.111$. So we know that 45% of 51 is 23.

35% of 60 is what number?

SOLUTION

35 is a percent, so it is put over 100. 60 is in the denominator because it is related to the OF in the equation. The unknown is in the numerator because it is related to the IS in the equation:

$$\frac{35}{100} = \frac{x}{60}$$

By cross multiplying, we get $100x = 2,100 \Rightarrow x = 2,100/100 \Rightarrow x = 21$. So 35% of 60 is 21.

 What percent of 48 is 36?

SOLUTION

Percent is the unknown here. The 36 is in the numerator, and 48 is in the denominator.

$$\frac{36}{48} = \frac{x}{100} \Rightarrow 48x = 3,600 \Rightarrow x = 3,600/48 \Rightarrow x = 75$$

Since 75 is a percent, 36 is 75% of 48.

 20 is what percent of 53?

SOLUTION

Again, the problem is asking for the percent, so the unknown is over 100. 20 is in the numerator, and 53 is in the denominator.

$$\frac{20}{53} = \frac{x}{100} \Rightarrow 53x = 2,000 \Rightarrow x = 2,000/53 \Rightarrow x = 37.74$$

By rounding off, we get 37.7. So 20 is 37.7% of 53.

 The original price of an air conditioner was $885.99. It was on sale, though, and sold for $708.79. What was the percent discount?

SOLUTION

For this problem, first we must subtract the sale price from the original price:

$885.99 − $708.79 = $1,77.20$ $1,77.20 is the amount of the discount

Then we must put it into the basic equation:

$$\text{Base (price)} \times \text{Rate} = \text{Percentage (tax amount)}$$

We must rearrange the basic equation above so that it gives us the rate. We must divide both sides by the base (price), so the equation looks like this:

$$\text{Rate} = \frac{\text{Percentage (tax amount)}}{\text{Base (price)}}$$

$$\frac{177.20}{885.99} = 20.00$$

Another way of solving this is to put it into a percent equation. The translated question is:

$$\$177.20 \text{ is what percent of } \$885.99?$$

Putting this into an equation, gives:

$$\frac{177.20}{885.99} = \frac{x}{100} \Rightarrow 885.99x = 17{,}720 \Rightarrow x = 17{,}720/885.99 \Rightarrow x = 20$$

So, the rate of the discount is 20%.

EXAMPLE 11

An MP3 player had a list price of $150.00. For a sale, the price was discounted 15%. Find the dollar amount of the discount and the sale price.

SOLUTION

Setting up the equation results in:

$$\text{Base} \times \text{Rate} = \text{Percentage (amount)}$$
$$150 \times 0.15 = P \Rightarrow P = 22.5$$

Alternatively, this can be translated directly into a percent problem.
What number is 15% of 150?

$$\frac{x}{150} = \frac{15}{100} \Rightarrow 100x = 2{,}250 \Rightarrow x = 2{,}250/100 \Rightarrow x = 22.5$$

So the amount of the discount is $22.50. The sale price (by subtracting $22.50 from $150.00) is $127.50.

Amal has a savings account that earns 3% annual interest. Last year, he made $75 in interest. How much did he originally have in the savings account?

SOLUTION

Setting up the equation gives:

$$\text{Base} \times \text{Rate} = \text{Percentage}$$
$$?\qquad .03\qquad \$75$$

We must rearrange the equation so that it calculates the base:

$$x = \$75/0.03 \Rightarrow x = 2{,}500$$

Alternatively, we could set up the percent proportion:

$75 is 3% of what number?

$$\frac{75}{x} = \frac{3}{100} \Rightarrow 3x = 7{,}500 \Rightarrow x = 7{,}500/3 \Rightarrow x = 2500$$

So the amount Amal had in the bank originally was $2,500.

DeLibero's had sales of $87,898 for the month of April. For the month of May, its sales were $75,450. What was the percent decrease from April to May?

SOLUTION

First, we need to calculate the amount of the decrease:

$$87{,}898 - 75{,}450 = 12{,}448$$

Setting up the equation gives:

$$\begin{array}{ccc}\text{Base} & \times\ \dfrac{\text{Rate}}{} = & \dfrac{\text{Percentage (amount)}}{}\\[4pt] 87{,}898 & ? & 12{,}448\end{array}$$

By solving for the rate, we have $x = 12{,}448 / 87{,}898 \Rightarrow x = 0.1416$. We round off to the nearest thousandth and change it into a percent: 14.2%.
Alternatively, we could set up the percent proportion:

$12,448 is what percent of $87,898?

$$\frac{12{,}448}{87{,}898} = \frac{x}{100} \Rightarrow 87{,}898x = 1{,}244{,}800 \Rightarrow x = 1{,}244{,}800/87{,}898 \Rightarrow$$
$$x = 14.16 \Rightarrow x = 14.2\%$$

So the percent decrease in sales for DeLibero's from April to May was 14.2%

EXERCISES

Exercise D   Solve the following problems, and check your answers:

28. 56 is what percent of 64?

29. 24 is what percent of 125?

30. What percent of 36 is 9?

31. What percent of 270 is 150?

32. 24 is what percent of 120?

33. 93 is what percent of 248?

34. What percent of 114 is 19?

35. What percent of 85 is 125?

36. $35 is what percent of $70?

37. $146 is what percent of $250?

38. What percent of $216 is $85?

39. $8 is what percent of $150?

40. $54.50 is what percent of $95?

41. What percent of $325 is $75?

In each of the following problems, find the percent of increase or decrease:

42. $315 decreased to $250

43. $128 increased to $325

44. $478 increased to $579

45. 18,240 decreased to 14,380

46. $515 increased to $735

47. $235 increased to $650

48. 6,434 increased to 8,420

49. $3,500 decreased to $2,300

Solve the following problems, rounding where necessary, and check your answers:

50. 35% of what amount is $72?

51. 45% of what amount is $395?

52. $369 is $33\frac{1}{3}$% of what amount?

53. 235 is 40% of what number?

54. 25% of what number is 2,460?

55. $248 is 110% of what amount?

56. 560 is 125% of what number?

57. 150% of what amount is $4,250.60?

58. 105% of what amount is $76.50?

59. 3,275 is 135% of what number?

Exercise E Find the missing items in the following forms, rounding where necessary:

60.

Sale Reduction Rates						
Original Price		Sale Price		Reduction Amount		Reduction Percent
$235	65	——	——	$ 82	48	——
330	33	$198	20	——	——	——
175	85	——	——	——	——	35%
575	95	316	77	——	——	——
425	75	——	——	140	50	——
380	25	197	73	——	——	——
298	95	194	32	——	——	——
605	85	——	——	——	——	28

61.

Totals Paid Including Sales Tax						
Amount Paid		Sales Tax Rate	Amount of Sale		Sales Tax Amount	
$ 85	78	6 %	——	——	——	——
128	65	——	$119	12	——	——
246	19	5.5	——	——	——	——
193	35	8.25	——	——	——	——
216	48	——	——	——	$10	31
342	31	4	——	——	——	——
259	17	——	248	01	——	——
342	20	6.5	——	——	——	——

Word Problems

The key phrases indicating a rate problem will not give you a numerical rate, but will ask for the rate or percent:

"What is the percent of . . . ?"
"What is the rate of . . .?"

The major difficulty lies in correctly identifying the base and the percentage. Read the problem carefully, analyze it logically, and then assign the correct values to both the base and the percentage.

Exercise F Solve the following problems, and check your answers:

62. Angel earns $975 a month. His rent is $175 a month. What percent of his monthly earnings does he spend on rent?

63. A piano regularly selling for $825 was reduced to $650. What is the percent of the reduction?

64. Ms. Donato earns $248 on a sale of $1,250. What is her rate of commission?

65. The Lang Department Store's sales for December were $68,750. Sales for January were $53,230. Find the percent of decrease for the month of January.

66. The average weekly food cost for a family of four is $78.50. Two years ago, the average weekly food cost for a family of four was $52.75. What is the percent of increase for the 2 years?

Percent problems in which you must find the base will always give you a rate and a percentage. It is essential to *correctly identify* each of the given numbers as the rate or the percentage. Read the problem carefully to make sure you understand what the question is asking for.

Exercise G Solve the following problems, rounding where necessary, and check your answers:

67. A retailer pays $125.50 for a dishwasher. If this will represent 65% of the selling price, find the selling price.

68. A retailer made $53,500 in net profits last year. If this represents 15% of his total sales, what were his total sales last year?

69. Rachel earns 25% more this year than last year. If she earns $16,575 this year, find the salary she earned last year.

Unit 3: Ratios and Proportions

RATIO

In the last unit, we did some proportions. In this section, we will discuss them in more detail, and we will start with ratios. A **ratio** is another name for a fraction. In other words, a ratio is a number over another number:

$$\frac{4}{5} \quad \frac{8}{9} \quad \frac{16}{3.4} \quad \frac{23.5}{6.75}$$

All of these are ratios. Note that the numbers could be decimal numbers, as we saw in the unit on percents. Suppose that in a store the ratio of female workers to male workers is 12 to 15. That is, there are 12 female workers and 15 male workers. The ratio can be shown in three ways:

$$\frac{12}{15} \quad 12 \text{ to } 15 \quad 12:15$$

Ratios can be shown in any of these three ways, depending on the context. We could reduce this fraction:

$$\frac{12}{15} = \frac{4}{5}$$

After reducing it, we could say that the ratio of females to males in the store is 4 to 5. This is a useful fact, and it is used all the time in business.

PROPORTIONS

A proportion is simply two equal fractions. Consider this:

$$\frac{1}{2} = \frac{x}{100}$$

What would you put in place of the x to make the fractions equal? Of course, the answer is 50 because 50/100 is equal to 1/2. This is exactly what a proportion is. Here are some other examples:

$$\frac{5}{8} = \frac{35}{56} \quad \frac{1}{4} = \frac{60}{240}$$

Setting up a proportion equation is usually the hardest part of solving a proportion. The following examples show how to set up proportions.

Carolann teaches nursing. Last year, she had 280 nursing students, and 6 out of 7 of the students were female. How many male and how many female students did she teach?

SOLUTION

The ratio is 6:7 and the total number of students is 280. The 7 in the ratio represents the total amount in the ratio, so we need to put that across from the total number of students in the proportion.

$$\frac{6}{7} = \frac{x}{280} \Rightarrow 7x = (280)(6) \Rightarrow 7x = 1,680 \Rightarrow x = 240$$
$$280 - 240 = 40$$

In the nursing classes, Carolann taught 240 females and 40 males.

 Roz's Roost serves ribs, and 3 out of 5 are served with hot, hot sauce. If Roz served up 411 orders of ribs having hot, hot sauce on them, how many orders of ribs did she sell? How many orders did she sell that did not have hot, hot sauce?

SOLUTION

The ratio is 3:5 and the total number of orders is 411. The 3 represents the number of orders having hot, hot sauce on them, so we place the 411 across from the 3 in the proportion:

$$\frac{3}{5} = \frac{411}{x} \Rightarrow 3x = (411)(5) \Rightarrow 3x = 2,055 \Rightarrow x = 685$$

The total number of orders of ribs was 685, and 685 – 411 = 274. 274 orders of ribs did not have hot, hot sauce on them.

 For every $7 of sales at the Soda Pop Shop, Henrietta, the owner, pays $4 in salaries. In one week, she had $7,126 in sales. How much did she pay in salaries?

SOLUTION

The ratio is 4:7, and the total number of sales is 7,126.

$$\frac{4}{7} = \frac{x}{7,126} \Rightarrow 7x = (7,126)(4) \Rightarrow 7x = 28,504 \Rightarrow x = 4,072$$

Henrietta paid $4,072 in salaries and had $3,054 left.

Gilda gets a commission of 14% of the gross sales of bicycles. In one week, she earned $859 in commissions. What gross amount did she sell in bicycles?

SOLUTION

The ratio is 14:100, and the commission is 859, so the 859 goes across from the 14 in the proportion, as shown.

$$\frac{14}{100} = \frac{859}{x} \Rightarrow 14x = (859)(100) \Rightarrow 14x = 85,900 \Rightarrow x = 6135.71$$

Gilda sold bicycles worth $6,135.71.

The Grants purchased a house and put down 11%. If they put down $19,250, what was the selling price of the house?

SOLUTION

The ratio is 11:100. The $19,250 is equal to the percent down, so it goes across from the 11 in the proportion:

$$\frac{11}{100} = \frac{19,250}{x} \Rightarrow 11x = (19,250)(100) \Rightarrow 11x = 1,925,000 \Rightarrow x = 175,000$$

The house the Grants bought cost $175,000.

Yuri's Appliance Store sells all sorts of appliances. The profit on each appliance is 40%. If Yuri's did $4,535 in sales one day, how much of this was profit?

SOLUTION

The ratio is 40:100. The $4,535 is the full amount of sales that day (100%). So the $4,535 goes across from the 100 in the proportion:

$$\frac{40}{100} = \frac{x}{4,535} \Rightarrow 100x = (4,535)(40) \Rightarrow 100x = 181,400 \Rightarrow x = 1,814$$

The profit that day at Yuri's Appliance store was $1,814.

The Planet Auto dealership sells Pluto automobiles. The profit on each car is 15%. If a car sells for $14,500, how much did the dealership make as profit?

SOLUTION

The ratio is 15:100. The $14,500 is the full price of the car (100%). So the $14,500 goes across from the 100 in the proportion:

$$\frac{15}{100} = \frac{x}{14,500} \Rightarrow 100x = (14,500)(15) \Rightarrow 100x = 217,500 \Rightarrow x = 2,175$$

The profit on a Pluto car is $2,175.

EXERCISES

Exercise A   Write each expression as a ratio in fraction form reduced to lowest terms.

1. 9 to 12
2. 15 to 5
3. 24:3
4. 12 out of 50
5. 3 shirts for $27
6. $3.60 for 3 pounds
7. $.10 per $1
8. $3.48 for 6 pounds

Word Problems

Exercise B   Solve the following problems, and check your answers:

9. A coat regularly selling for $160 is on sale at $\frac{1}{4}$ off. What is the amount of the reduction?

10. A clothing store had 288 pairs of slacks in stock. If $\frac{3}{8}$ of the slacks were sold, how many pairs of slacks were sold?

Exercise C   Find the missing quantities. Check each answer.

11. $\frac{35}{14} = \frac{x}{2}$
12. $\frac{x}{48} = \frac{5}{8}$
13. $x:6 = 40:48$
14. $5:x = 45:63$

15. 2 is to 13 as x is to 1.105
16. $5:11 = x:693$
17. 6 oz.:1 lb. $= x:$ 592 lb.

18. 15% of $235
19. 25% of what number is 130?

Word Problems

Exercise D   Solve the following problems, and check your answers:

20. Jane is paid a commission of 12% of total sales. If she earned $393 in commissions, what were her total sales?

21. An appliance store sells a refrigerator for $625. If the gross profit is 40%, what is the amount of the gross profit?

Unit 4: Review of Chapter 2

Change the following common fractions to decimals to the nearest *tenth*:

1. $\dfrac{4}{5}$ 2. $\dfrac{14}{17}$

Change the following common fractions to decimals to the nearest *hundredth*:

3. $\dfrac{1}{3}$ 4. $\dfrac{18}{30}$

Change the following common fractions to decimals to the nearest *thousandth*:

5. $\dfrac{14}{32}$ 6. $\dfrac{23}{36}$

Solve the following problem:

7. Cynthia's take-home pay is $\dfrac{2}{3}$ of her gross income of $1,500 a month. If she saves $\dfrac{1}{8}$ of her take-home pay, how much does she save each month?

8.

		Wholesaler's Trade Discounts				
		Effective January 1–July 1				
Item Code	List Price	Trade Discount	Amount of Discount		Net Price	
A	$127	40%	____	__	____	__
B	163	45	____	__	____	__
C	272	55	____	__	____	__
D	245	35	____	__	____	__
E	286	43	____	__	____	__
F	342	52	____	__	____	__
G	678	47	____	__	____	__

Find the amount of interest and the new balance for each of the following amounts. Round off each number to the nearest cent.

9.

Amount on Deposit		Interest Rate	Amount of Interest		New Balance	
$1,238	65	8.752%	———	——	———	——
1,578	42	9.75	———	——	———	——
2,415	70	8.025	———	——	———	——
2,571	81	9.65	———	——	———	——
3,682	55	10.0325	———	——	———	——
3,520	60	11.652	———	——	———	——

Find the amount to the nearest penny or whole unit.

10. $322 is 115% of what amount?

11. $462.75 is 112% of what amount?

12. Janet Gomez earns a 10% commission on all her sales. In January, her sales were $53,694.75, and in February her sales were $42,785.50.

    (a) How much less commission did she earn in February?

    (b) What was the percent of decrease in her earnings for February as compared to January?

13. Sam Grotin earned $728.50 in interest on his saving account last year. If the rate of interest is $6\frac{1}{2}$%, how much did Sam have on deposit at the beginning of last year?

14. At the end of a full year Mary Garber had $6,248.95 in her savings account. If the rate of interest was 12.253%, how much money did Mary have in the bank at the beginning of the year?

Statistics, Charts, and Graphs

Unit 1: Measures of Central Tendency

There are several measures of central tendency. Most people are familiar with the average, also called the mean. Two other measures of central tendency, median and mode, will be discussed in this section.

THE MEAN (AVERAGE)

The mean (average) is calculated by adding up all the numbers in the set under consideration and dividing by the quantity of numbers.

 What is the average of these numbers?

$$23, 57, 45, 29, 31, 52, 23, 55, 38, 36$$

SOLUTION

There are 10 numbers. Add them:

$$23 + 57 + 45 + 29 + 31 + 52 + 23 + 55 + 38 + 36 = 389$$

Divide the sum by the quantity of numbers:

$$389/10 = 38.9$$

The mean of the numbers is 38.9.

 Seven waiters and waitresses agreed to pool their tips at the Short Stop Diner and take home the average:

$$\$45, \$62, \$72, \$56, \$49, \$54, \$68$$

How much did each waiter and waitress earn?

SOLUTION

$$45 + 62 + 72 + 56 + 49 + 54 + 68 = 406$$
$$406/7 = 58$$

So each waiter and waitress earns $58 that day.

THE MEDIAN

The median is the middle number of a group of numbers. To find the median, it is easiest to order (sort) the numbers from lowest to highest (or from highest to lowest).

What is the median of these nine numbers?

$$3, 6, 2, 7, 9, 1, 6, 12, 2$$

SOLUTION

Sorting these numbers gives:

$$1, 2, 2, 3, 6, 6, 7, 9, 12$$

The first 6 is the middle number, so that is the median.

Finding the median is easy when the quantity of numbers is odd. We simply take the middle number. What happens, though, when we have an even quantity of numbers?

What is the median of these numbers?

$$16, 12, 15, 21, 11, 12, 16, 19$$

SOLUTION

Sorting these numbers gives:

$$11, 12, 12, 15, 16, 16, 19, 21$$

We see that the middle two numbers are 15 and 16. So take the average of these two numbers to find the median.

$$15 + 16 = 31 \quad 31/2 = 15.5$$

The median of these numbers is 15.5.

One of the most common uses of the median is with home prices in a town or in a section of a town. The median home price is used much more often than the average home price by real estate agents.

THE MODE

The mode is the most frequently occurring number in a group of numbers. As with the median, it is usually easiest to sort the numbers first. Then the mode(s) show themselves clearly.

EXAMPLE 5   What is the mode of these eight numbers?

$$23, 12, 8, 16, 14, 5, 8, 22$$

SOLUTION

First, we sort them:

$$5, 8, 8, 12, 14, 16, 22, 23$$

Then we look for the most frequently occurring number, which is 8. So 8 is the mode. What happens when we have two frequently occurring numbers?

 EXAMPLE 6   We are going to use the same set as in Example 4: 16, 12, 15, 21, 11, 12, 16, 19. What is the mode?

SOLUTION

First, we sort them:

$$11, 12, 12, 15, 16, 16, 19, 21$$

In this set, two numbers occur twice, so we have *two* modes, 12 and 16. There can be a great many modes, but it is always the most frequently occurring number in the set under consideration.

 EXAMPLE 7   What is (are) the modes in this set?

$$4, 17, 25, 32, 3, 33, 4, 32, 24, 14, 29, 17, 5, 9, 17, 27, 33, 32, 19, 8$$

SOLUTION

First, we sort the set:

Hint: With large sets of numbers, it is easy to miss one or two. To avoid this, try counting the numbers once they have been sorted. (There are 20 numbers here).

$$3, 4, 4, 5, 8, 9, 14, 17, 17, 17, 19, 24, 25, 27, 29, 32, 32, 32, 33, 33$$

By scanning the sorted numbers, we can see that there are again two modes: 17 and 32. They both occur three times in the set.

USING A CALCULATOR

Using a calculator makes doing most of these calculations quite easy. In fact, some calculators have an average function that will calculate the mean (average) for you. Entering all the values in a set of numbers and pressing the "average" key will result in the average being displayed on the screen.

COMPARISON OF MEASURES OF CENTRAL TENDENCY

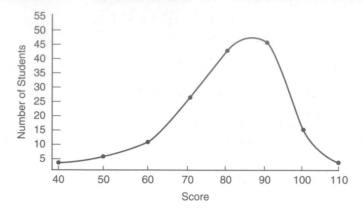

On a recent test of 10 questions plus a bonus question, 171 members of the freshman class received these marks:

Score	Number of Students
40%	2
50	5
60	12
70	30
80	48
90	52
100	18
110	4

HINTS: To find the *average*, take 2 × 0.40 plus 5 × 0.50, etc., and then divide by 171. This gives an average of

$$\frac{139.10}{171} = 0.81345 = 81.35\%$$

To find the *median*, count in 86 people from the top and from the bottom. The score of the person right in the middle is the median score. In this case, the median score is 80%.

To find the *mode,* the score that occurs most frequently, count the number of times each number occurs. In this case, the mode is 90%.

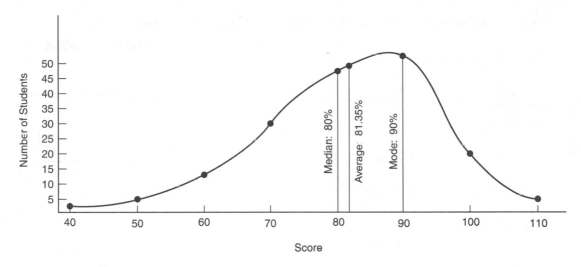

Notice that of the above three measures only the *median* (in a set with an odd number of items) and the *mode* are items that are actually contained in the given sets of numbers. When deciding which of the three measures you should use with a given set of statistics, you must determine the purpose of the information:

- If you need a number that is typical of the set, the *mean* can be used, unless a large number of high or low scores cause the mean to be nonrepresentative.
- If you need a number about which the others are evenly spaced, the *median* can be used.
- If you need a number that is more prominent because of its frequency, the *mode* can be used.

EXERCISES

Exercise A   Find the *mean* for each set of numbers:

1. 20, 28, 45, 25, 34, 40

2. 325, 340, 315, 330, 310, 305

3. 260, 242, 251, 263, 250

4. 460, 443, 455, 470, 462

5. 18, 22, 16, 19, 20, 17, 21, 19, 24, 25

Find the *median* and the *mode* in each of the following sets of numbers:

	Median	Mode
6. 70, 85, 80, 65, 70	_____	_____
7. 24, 32, 24, 20, 32, 25, 32	_____	_____
8. 191, 152, 183, 191, 140, 170, 164	_____	_____
9. 42, 53, 40, 61, 40, 64, 58, 40	_____	_____
10. 195, 170, 180, 190, 165, 170, 185	_____	_____

Word Problems

Exercise B   Solve the following problems:

11. A salesperson made the following sales last week: Monday, $248.50; Tuesday, $310.78; Wednesday, $278.65; Thursday, $320.43; Friday, $315.15; and Saturday, $343.85. What is the daily average of the sales made?

12. An office employs six clerks at the following salaries: $9,850; $8,975; $11,475; $10,500; $12,250; and $9,750.

    (a) What is their mean annual salary?
    (b) What is their median salary?

13. The average weekly salary of six salesclerks is $195. Five of the salesclerks earn the following salaries: $185, $190, $205, $195, and $200. What is the salary of the sixth salesclerk? (HINT: Since the average salary is $195, the total weekly salaries amount to $195 $\times$ 6, or $1,170.)

Unit 2: Line Graphs

The Graduate Management Admission Test (GMAT) is the entrance exam into graduate studies in business management. A significant part of the test is concerned with the interpretation of graphs and charts. They are the key to making sense of the direction of the market. Tables, or columns, of numbers can be confusing. Graphs and charts can show trends in the market much more clearly.

A line graph is a two-dimensional representation of a set of numbers. It is, quite literally, a picture of a table of two related numbers. In business, many quantities and relationships are shown with line graphs. In the following example, we are looking at the amount of sales for a year.

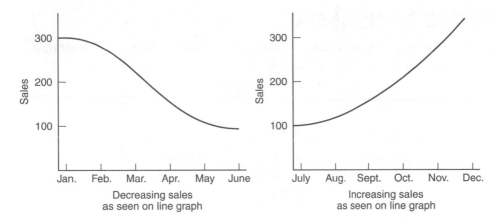

Decreasing sales
as seen on line graph

Increasing sales
as seen on line graph

Note that there is a vertical scale (the sales) and a horizontal scale (the months of the year). This will always be the case with linear graphs. In the first graph, the sales went down in the first half of the year (the sales were decreasing). In the second graph, the sales went up in the second half of the year (the sales were increasing).

Let's look at a simple example. Kelly tracked the afternoon sales in her department (the sewing dept.) of Hahn's Department Store. Here is the hour-by-hour sales:

Hour	Sales
1	$45
2	$63
3	$55
4	$65
5	$48

Here is the graph of this data:

Sales at Hahn's Sewing Dept.

This graph was done on a spreadsheet.

PREPARING A LINE GRAPH

To draw a line graph, you need a table comparing one quantity, like sales, to another quantity, like months. Here is a set of data to use:

Appliance Department Sales for 2007

Month	Sales	Sales to the Nearest $1,000
January	$15,468.95	$15,000
February	18,763.50	19,000
March	20,394.75	20,000
April	16,643.25	17,000
May	25,415.30	25,000
June	23,146.90	23,000
July	34,861.40	35,000
August	37,419.70	37,000
September	24,718.20	25,000
October	21,953.75	22,000
November	32,385.70	32,000
December	27,410.90	27,000

To draw a graph, follow these steps:

1. Get a piece of graph paper.

2. Prepare the data in a table, such as the one above.

3. Round off numbers to the nearest $1,000, as the table shows.

4. Mark the vertical axis in even increments, in this case, thousands of dollars.

5. Mark the horizontal axis in even increments, in this case, months of the year.

6. Plot the points corresponding to the proper dollar amount for the proper month.

7. Connect the points, and check the graph for possible mistakes.

Hint: There is no need to start at zero for either the horizontal or the vertical axes. In fact, sometimes the data are clearer if the graph starts at 1,000 or at 100,000.

The graph for this table is in the exercises at the end of this section. You should draw it first and then compare it to the one in the book.

CALCULATOR OPTION

Another way to draw a line graph is to use a calculator. Several brands and models allow for drawing graphs from a table of values. Most graphing calculators have a "table" function that allows for the input of values into a table. The calculator will have a "graph" function that will take that data and graph it. Be sure to look for the prompt to choose a line graph from among the graphing options.

SPREADSHEET OPTION

Still another way to draw a line graph is to use a computer spreadsheet. As with the calculator, values can be entered into the spreadsheet and a graph function in the spreadsheet will draw a line graph. To prepare a line graph on a spreadsheet, follow these steps:

1. Open up a spreadsheet.

2. Enter the two sets of data in two columns of the spreadsheet (not rows). Do not enter any other data (in other columns) so that the spreadsheet will know what columns to choose for the graph.

3. Choose the "chart wizard" function from the drop-down menu.

4. Follow the prompts to choose a line graph and to label both the *x*- and the *y*- coordinates.

When you press "finish" in the chart wizard, the graph should appear in your spreadsheet. The first graph in this section (Hahn's Sewing Dept.) was made with a spreadsheet.

EXERCISES

Exercise A   Examine the line graph below, and answer the questions that follow.

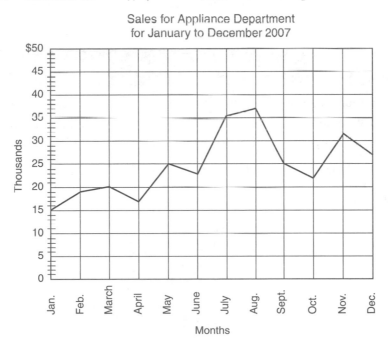

Sales for Appliance Department
for January to December 2007

1. What information does the graph represent?

2. How many $1,000 in sales is represented by each major division on the vertical scale?

3. (a) In what month were the sales the highest?
   (b) How much was sold in that month?

4. In what two months was the amount of sales the same?

5. Which months showed increases in sales over the previous month?

6. (a) Which month had the greatest increase in sales over the previous month?
   (b) What were the sales of that month?

7. In what month were the sales $22,000?

8. What month shows an increase in sales of $12,000?

Exercise B   Prepare line graphs for the following two groups of data:

9. The Photo-Tech Manufacturing Company introduced a dry-method photo-copier in 1997. The sales for 1998–2007 were as follows: 1998, $52,365; 1999, $87,872; 2000, $138,410; 2001, $211,865; 2002, $263,635; 2003, $315,880; 2004, $290,235; 2005, $364,415; 2006, $415,605; 2007, $475,675. Let the horizontal scale represent the 10-year period from 1998 to 2007 and the vertical scale represent the yearly sales. Let each major division on the vertical scale represent $50,000 in sales and each subdivision represent $10,000 in sales. Round off the given amounts to the nearest $10,000.

Unit 3: Stem-and-Leaf Plots

A stem-and-leaf plot is a type of graph that has gotten more popular recently. The big advantage of a stem-and-leaf plot is that none of the data is lost; all of it is in the plot. Suppose we have the ages of the 15 workers at Donovan's Paint Company:

34, 55, 32, 28, 56, 25, 63, 54, 59, 44, 32, 31, 48, 43, 25

Making a stem-and-leaf plot for this data is easy:

2 | 5, 5, 8
3 | 1, 2, 2, 4
4 | 3, 4, 8
5 | 4, 5, 6, 9
6 | 3

Note that all the data is available; it is just in a more compact form. The numbers are sorted. We can see the median and the mode from this rather easily (median: 43; modes: 25 and 32).

EXERCISES

---

1. Greg's restaurant has ten waiters and waitresses. At the end of one night, they compared their tips. The amounts they earned were:

   $45, $67, $55, $53, $65, $55, $48, $50, $52, $58

   Construct a stem-and-leaf plot for this data. Then find the median and mode.
   Note: The dollar sign is generally not needed when making a stem-and-leaf plot.

2. Uncle John's Pancake House had different numbers of tables occupied each of the ten hours they were open one day. Here are the numbers of tables occupied:

   12, 15, 18, 21, 25, 23, 26, 18, 15, 13

   Construct a stem-and-leaf plot for this data. Then find the median and mode.

3. The scouts were selling holiday wreaths. A large wreath sold for $15, and a small wreath (spray) sold for $7. The twelve scouts who sold wreaths came back with these total sales:

   $15, $21, $45, $30, $37, $22, $44, $37, $45, $22, $28, $37

   Construct a stem-and-leaf plot for this data. Then find the median and mode.

4. At a craft fair, the fifteen vendors tallied up their gross sales at the end of one day:

   $53, $55, $96, $75, $83, $94, $74, $85, $59, $84, $83, $78, $66, $73, $85

   Construct a stem-and-leaf plot for this data. Then find the median and mode.

5. Tom's Toasted Peanut Shop, down on the Jersey Shore, tallied up the sales in the last two-week period. Here are those totals:

   $89, $69, $95, $85, $91, $69, $77, $81, $89, $95, $75, $89

   Construct a stem-and-leaf plot for this data. Then find the median and mode.

Unit 4: Bar Graphs and Pictographs

A **bar graph** is a graph that shows data in the form of bars, double bars, or 3-D bars. Data sometimes is easier to understand when shown as bars rather than lines. A **pictograph** is similar, but it uses little pictures of the data being displayed instead of bars. For instance, if the pictograph displays groceries sold at a supermarket, the data might look like grocery bags (with each grocery bag representing a set amount of sales).

Bar graphs and pictographs are designed to compare values over time or across departments in a department store, for instance. The heights of the bars or the number of pictures clearly show the difference between departments, across time, or whatever the quantities being compared are. These graphs can be vertical or horizontal. Consider the following bar graph:

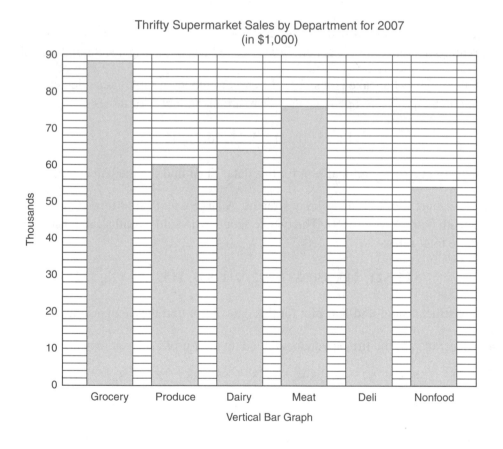

Thrifty Supermarket Sales by Department for 2007
(in $1,000)

This bar graph could be easily displayed as horizontal bars, with the horizontal and vertical axes interchanged. The data table that generated this bar graph is below:

**Thrifty Supermarket Sales by Department
for 2007**

Department	Sales	Sales to the Nearest $1,000
Grocery	$88,465.90	$88,000
Produce	59,864.75	60,000
Dairy	63,593.50	64,000
Meat	76,253.48	76,000
Deli	42,480.70	42,000
Nonfood	53,505.80	54,000

PREPARING A BAR GRAPH

The steps in preparing a bar graph are similar to those in preparing a line graph. To draw a graph, follow these steps:

1. Get a piece of graph paper.

2. Prepare the data in a table, such as the one above.

3. Round off numbers to the nearest $1,000, as the table shows.

4. Mark the vertical axis in even increments, in this case, thousands of dollars.

5. Mark the horizontal axis in even increments, in this case, store departments (grocery, produce, and so on).

6. Place a dot corresponding to the proper dollar amount for the proper department.

7. Draw a bar corresponding to that point.

8. Continue to plot points and draw bars for all the data.

9. Check the graph for possible mistakes.

HINT: As with line graphs, there is no need to start at zero for either the horizontal or the vertical axes. Indeed, sometimes the data is clearer if the graph starts at 1,000 or at 100,000.

CALCULATOR OPTION

Another way to draw a bar graph is to use a calculator. Several brands and models allow for drawing graphs from a table of values. Most graphing calculators have a "table" function that allows for the input of values into a table. The calculator will have a "graph" function that will take that data and graph it. Be careful to look for the prompt to choose a bar graph from among the graphing options.

SPREADSHEET OPTION

Still another way to draw a bar graph is to use a computer spreadsheet. As with the calculator, values can be entered into the spreadsheet and a graph function in the spreadsheet will draw a bar graph. To prepare a bar graph on a spreadsheet, follow these steps:

1. Open up a spreadsheet.
2. Enter the two sets of data in two columns of the spreadsheet (not rows). As with the line graph, do not enter any other data (in other columns) so that the spreadsheet will know what columns to choose for the graph.
3. Choose the "chart wizard" function from the drop-down menu.
4. Follow the prompts to choose a bar graph and to label both the *x*- and the *y*- coordinates.
5. There are several options to choose from in the bar graph option: bar, split bar, 3-D bar, and so on. Choose the one that suits your data.

EXAMPLE 1   Dave has a tropical fish business. One day, he looked at his inventory and found these species and amounts:

DAVE'S TROPICAL FISH

Name	Amount
Guppies	8
Placos	5
Swordtails	7
Mollies	9
Tetras	6

Create a bar graph from this data.

SOLUTION

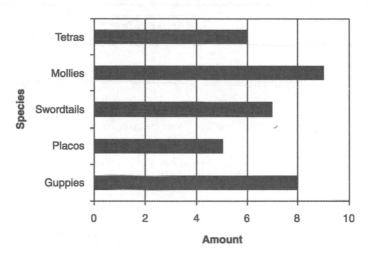

Note: This is an example of a bar graph done with an electronic spreadsheet.

PREPARING A PICTOGRAPH

The steps in preparing a pictograph are identical to those in preparing a bar graph, except that a picture is chosen to represent the quantity being considered. Generally, a picture is used to show a whole amount, such as 1,000 (vehicles, as in the following example). If the data shows a half amount (500, if the whole amount was 1,000), then the picture will be of half a car. Thus, pictographs can easily show precision to about 1/4 (showing a quarter of a car, half of a car, or three-quarters of a car).

Yearly Car Sales for the Ace Car Company
for 1995 to 2000

Year	Number of Cars Sold
1995	355
1996	490
1997	445
1998	653
1999	624
2000	853

The pictograph for this table has been done in the exercises at the end of this section. However, try making this one up before looking at it in the exercises to see how you do preparing a pictograph.

EXERCISES

Exercise A   Examine the horizontal bar graph below and answer the questions that follow.

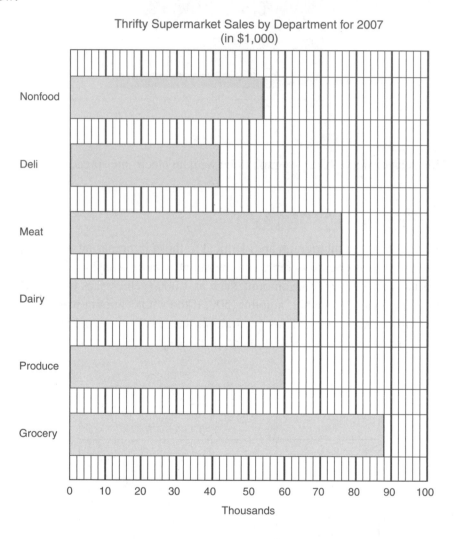

Thrifty Supermarket Sales by Department for 2007
(in $1,000)

1. What information does the graph represent?

2. How many $1,000 in sales is represented by each major division on the horizontal scale?

3. (a) What department had the highest sales?
   (b) What were the sales of that department?

4. What were the sales of the deli department?

5. What departments had sales of under $60,000 each?

6. Which two departments show the greatest difference in sales?

7. What percentage of the total sales is the sales of the grocery department?

8. List the departments in descending order according to sales.
   (a) Which department sold $42,000?
   (b) Which two departments show a difference of $24,000?

Exercise B   Examine the horizontal pictograph below and answer the questions that follow.

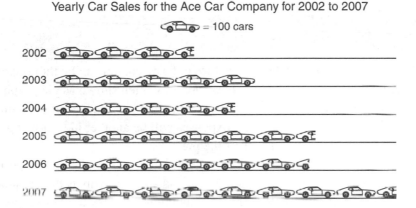

Yearly Car Sales for the Ace Car Company for 2002 to 2007

= 100 cars

9. Which year had the lowest sales?

10. How many cars were sold in 2004?

11. How many cars were sold in 2002?

12. What is the percent of increase in sales in 2007 over the sales in 2006?

Word Problems

Exercise C   Prepare the following graphs:

13. Prepare a vertical bar graph for the following data:

   The Acorn Manufacturing Company sold the following amounts in its five territories last month: territory A, $47,625.80; B, $32,249.75; C, $51,817.45; D, $26,485.75; E, $29,525.35.

Let the horizontal scale represent the five territories and the vertical scale represent the amounts of monthly sales. Let each major division on the vertical scale represent $10,000 in sales and each subdivision represent $2,000 in sales. Round the given amounts to the nearest $1,000.

14. Prepare a pictograph for the following data:

The average tuition costs for 1 year at a state college during 2003 to 2007 were as follows:

2003, $5,175; 2004, $5,768; 2005, $6,448; 2006, $6,848; 2007, $7,281. Round off each tuition cost to the nearest $100. Use the picture of a $1,000 bill,

, to represent $1,000.

Unit 5: Circle (Pie) Graphs

The circle (pie) graph is used when the values to be discussed can be thought of as a whole. Budgets (both income and expenses) are very often shown as pie graphs. The income will be shown as a pie, with pieces showing the percentages of each individual income stream as pieces of the pie. The expenses will be another pie, showing all the individual streams going out as pieces of the pie.

In the following pie graph, it is easy to see that category A is the largest portion of the graph, that category B is the second largest, and so forth. This is the great advantage of the pie graph. It represents 100% of a quantity. It is easy to see *at a glance* which quantities are the most significant and which ones are the least significant. The sizes of the pie slices show the importance of the different categories.

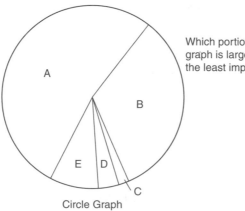

Which portion of this circle graph is largest? Which is the least important?

Circle Graph

PREPARING PIE GRAPHS

Unlike the line or bar graphs, we do not need a table with two quantities to draw a pie graph. We need a table with only one quantity, like sales, income, or expenditures. Here is a set of data to use:

Food	$ 5,700	20%
Rent	6,270	22
Savings	4,275	15
Clothing	2,280	8
Recreation	4,845	17
Health	1,425	5
Charity	1,140	4
Miscellaneous	2,565	9
	$28,500	100%

Steps to prepare a pie chart:

1. Add up the individual amounts to get the total, as shown.

2. Calculate the percent of the total for each individual amount. Use the percent equation from Chapter 2. Let's do one to illustrate. What percent of $28,500 is $5,700 (the food)? This translates into the equation:

$$\frac{x}{100} = \frac{5,700}{28,500} \Rightarrow 28,500x = 570,000 \Rightarrow x = 20 \Rightarrow x = 20\%$$

Do all the other expenses in the same manner.

3. Convert all the percentages into slices of the pie. That is, convert them to angle measures. The circle is 360°. 20% of 360 is:

$$\frac{20}{100} = \frac{x}{360} \Rightarrow 100x = 7,200 \Rightarrow x = 72°$$

4. Continue with all the other quantities until all the expenses are in degree format. They should add up to 360°. You may have to add or subtract a degree or two to make it work out.

5. Draw a circle (of around 1-inch radius or 2-inch diameter), and get a protractor. Use the protractor to measure the angles for all the pieces. For example, measure a 72° angle for the food.

6. Check the pie graph for possible errors.

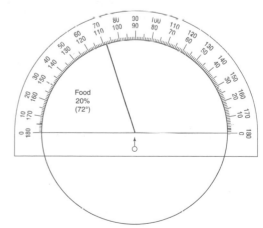

CALCULATOR OPTION

No calculators will construct pie graphs.

SPREADSHEET OPTION

Another way to draw a pie graph is to use a computer spreadsheet. As with the calculator, values can be entered into the spreadsheet and a graph function in the spreadsheet will draw a pie graph. To prepare a pie graph on a spreadsheet, follow these steps:

1. Open up a spreadsheet

2. Enter the one set of data in one column of the spreadsheet. As with the line and bar graphs, do not enter any other data (in other columns) so that the spreadsheet will know what columns to choose for the graph.

3. Choose the "chart wizard" function from the drop-down menu.

4. Follow the prompts to choose a pie graph and to label both the x- and the y- coordinates.

5. There are several options to choose from in the pie graph option: pie, 3-D pie, and so on. Choose the one that suits your data.

John has a collection of beer mugs. Here is his collection:

JOHN'S BEER MUGS

Type	Number
Sam Adams	6
Guinness	12
Genesee	2
Grant's	3
Budweiser	1

Create a pie graph from this data.

SOLUTION

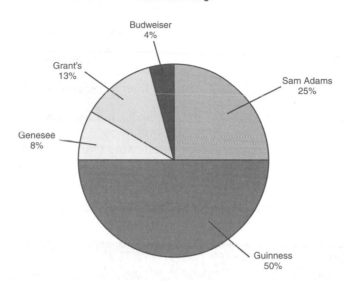

John's Beer Mugs

Budweiser 4%

Grant's 13%

Sam Adams 25%

Genesee 8%

Guinness 50%

Note: This is an example of a pie graph done with an electronic spreadsheet.

EXERCISES

Exercise A   Examine the circle graph below and answer the questions that follow.

Distribution of the Fernandez Annual Income of $28,500

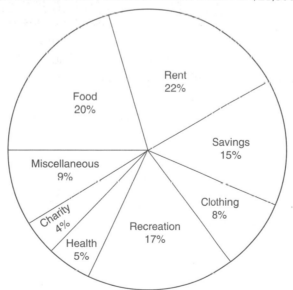

Rent 22%

Food 20%

Savings 15%

Miscellaneous 9%

Clothing 8%

Charity 4%

Recreation 17%

Health 5%

1. What information does the graph represent?

2. What should be the total percent of all the sectors added together?

3. What expenditure is represented by the smallest sector of the circle?

4. What is the total percent of the three largest expenditures?

5. How much money does the Fernandez family spend on rent?

6. How much more money does the Fernandez family spend on rent than on food?

Exercise B   Draw circle graphs for the following two groups of data:

7. The breakdown in sales of the Bee Gee Appliance Company for last year was as follows: refrigerators, $178,464; washers, $118,976; dryers, $75,712; dishwashers, $54,080; television sets, $81,120; stereos, $32,448.

Unit 6: Review of Chapter 3

SUMMARY OF KEY POINTS

• Mean (average) $= \dfrac{\text{Sum of the scores}}{\text{Number of scores}}$ The mean is the best all-around measure of central tendency, but gives an answer that does not relate to any one real score.

• Median is the *middle* score of a set.   Most times, the median gives an actual score. It also tends to minimize the effect of a few abnormally high or low scores.

• Mode is the *single most popular* score.   The mode is an actual score, being the single most popular score. The mode is immune to extremes.

• Line graphs should be used to demonstrate changes and trends over time.
• Bar graphs or pictographs should be used to show comparisons between groups in *real numbers* (dollars, units, etc.).
• Circle graphs should be used to show comparisons between groups in *relative* terms (percent).

Find the mean in each of the following sets of numbers:

1. 35, 45, 30, 43, 32, 48, 33

2. 320, 348, 372, 328, 350, 365

Find the *median* and the *mode* in each of the following sets of numbers:

	Median	Mode

3. 85, 70, 95, 80, 75, 70, 75, 70 _____ _____

Examine the line graph below, and answer the questions that follow.

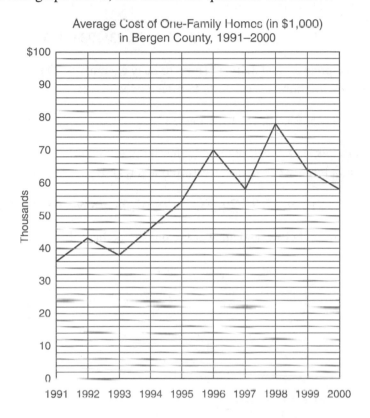

4. (a) In what year was the cost the highest?
   (b) How much was the cost that year?

5. What year showed the largest increase from the preceding year?

6. Arrange all the yearly prices as a set. Find the mean, median, and mode of the prices represented by the graph.

For each of these exercises, draw a stem-and-leaf plot, and then calculate the mean, median, and mode.

7. The Watchung Booksellers looked at sales over a three-week period (fifteen days). The sales were:

$553, $455, $566, $423, $654, $456, $429, $562, $552,
$657, $632, $425, $454, $654, $556,

8. Grove Pharmacy filled prescriptions in each of the thirteen hours they were open one day. Here are the numbers of prescriptions filled:

   12, 35, 28, 29, 35, 37, 56, 68, 55, 53, 39, 52, 25

9. Tomasso's Terrific Tacos and Tamales tallied the numbers of items they sold in a ten-day period. Here are the tallies:

   150, 230, 175, 155, 235, 227, 176, 235, 233, 175

10. Karen cleans, presses, and delivers shirts to busy executives. One day, she delivered the following numbers of shirts to fifteen executives:

    19, 25, 23, 12, 15, 13, 14, 25, 13, 14, 22, 18, 16, 13, 15

11. Kim loves to talk on her cell phone, but her mom tallied up the minutes per day she talked in a three-week period (twenty-one days). Here are those totals:

   99, 67, 85, 75, 91, 79, 67, 91, 79, 93, 75, 98, 67, 71, 78, 92, 68, 74, 81, 79, 67

Examine the bar graph below, and answer the questions that follow.

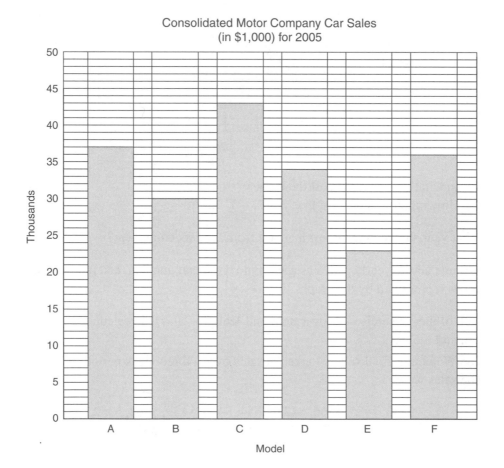

Consolidated Motor Company Car Sales
(in $1,000) for 2005

12. What is the difference in the number of cars sold between the model that had the lowest sales and the model that had the highest sales?

13. What is the model C sales percent of the total cars sold?

Examine the circle graph below, and answer the questions that follow.

Dollar Distribution of Gross Sales
Tillary Department Store

Cost of merchandise
60%

Net profit
5%

Salaries
14%

Maintenance
6%

6%

Rent
9%

14. What two expenditures are equal to the salary expenditure?

15. If the gross sales are $1,328,467, find the expenditure for each item in the graph.

16. If the net profit for a year is $71,157.85, what would be the cost of merchandise for the year?

Examine the pictograph below, and answer the questions that follow.

Two-Phone Households in Winston County, 2000–2004

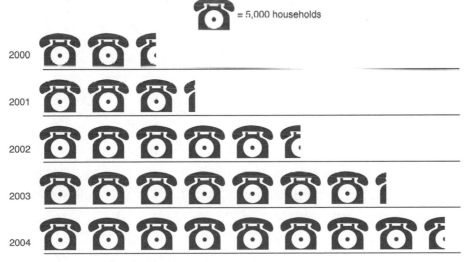

= 5,000 households

2000

2001

2002

2003

2004

17. How many more two-phone households were there in 2004 than in 2000?

Measurement– English (U.S. Standard)/ Metric Conversion

The United States has, historically, always used the English (the customary) weights and measures system (inch, foot, ounce, pound, quart, gallon, and so on). Most of the world, however, uses the metric system. Increasingly, this country is moving to that system. A complete change may be a decade or more away, so it is important to know how to change from English to metric and back again. Fortunately, the conversions are not too difficult. As with some other topics, many calculators have a function that converts from the English system to the metric system.

Unit 1: Converting Standard Units

In many business situations, you will need to convert inches to feet, feet to yards, (inches to yards), pounds to ounces, ounces to gallons, and so on. Just be careful, and you will not have any problems. The following table shows the conversions.

TABLE 1
UNITS OF MEASURE

Length

1 foot (ft.) − 12 inches
1 yard (yd.) = 3 feet or 36 inches
1 mile (mi.) = 5,280 feet or 1,760 yards

Capacity

1 pint (pt.) = 16 liquid ounces
1 quart (qt.) = 2 pints or 32 liquid ounces
1 gallon (gal.) = 4 quarts or 128 liquid ounces
    = 8 pints

Weight

1 pound (lb.) = 16 ounces
1 short ton (s.t. or T) = 2,000 pounds
1 long ton (l.t.) = 2,240 pounds

Quantity

1 dozen (doz.) = 12 units
1 gross (gr.) = 12 dozen or 144 units
1 score = 20 units

EXAMPLE 1   Change 108 inches to feet and to yards.

SOLUTION

A proportion will work here. 12 inches is to 1 foot as 108 inches is to how many feet?

$$\frac{12}{1} = \frac{108}{x} \Rightarrow 12x = 108 \Rightarrow x = 9$$

108 inches equals 9 feet.

Another proportion will work for converting into yards. 36 inches is to 1 yard as 108 inches is to how many yards?

$$\frac{36}{1} = \frac{108}{x} \Rightarrow 36x = 108 \Rightarrow x = 3$$

108 inches equals 3 yards.

Of course, once we converted from inches to feet, we could have converted from feet to yards.

Alternatively, we could consider how many feet (12-inch groups) can be taken out of 108? Since $9 \times 12 = 108$, we have 9 feet in 108. Then figure out how many yards (3-feet groups) are in 9 feet? The answer is 3. Knowing the multiplication tables is really helpful.

 EXAMPLE 2   Change 123 inches to feet and to yards.

SOLUTION

A proportion will work here: 12 inches is to 1 foot as 123 inches is to how many feet?

$$\frac{12}{1} = \frac{123}{x} \Rightarrow 12x = 123 \Rightarrow x = 10.25$$

We want feet and inches. Separate the 10 feet, which leaves 0.25 feet left over. Since $12 \times 0.25 = 3$, 123 inches equals 10 feet 3 inches.

Starting with this result, we convert to yards. There are 3 feet in a yard, so we have:

$$10 \text{ feet } 3 \text{ inches} \Rightarrow 10/3 = 3 \text{ yards, } 1 \text{ foot.}$$

So 123 inches equals 3 yards 1 foot 3 inches.

EXAMPLE 3

Change 86 ounces into pounds.

SOLUTION

The proportion is 16 ounces is to 1 pound as 86 ounces is to how many pounds?

$$\frac{16}{1} = \frac{86}{x} \Rightarrow 16x = 86 \Rightarrow x = 5.375$$

There are 5 full pounds. However, we want to know how many pounds and ounces there are. After separating the 5 pounds, we have 0.375 pounds left. Multiply this by 16 to get the ounces:

$$0.375 \times 16 = 6 \text{ ounces.}$$

So 86 ounces equals 5 pounds 6 ounces.

Alternatively, we could consider how many groups of 16 are in 86. By dividing, we have:

$$16\overline{)86} \\ \quad \underline{80} \\ \quad\; 6 \quad\; \overset{5}{}$$

So by long division, we have 5 with a remainder of 6. So the answer is 5 pounds 6 ounces.

EXAMPLE 4

Change 150 liquid ounces into gallons, quarts, and liquid ounces.

SOLUTION

The proportion is 32 liquid ounces is to 1 quart as 150 liquid ounces is to how many quarts?

$$\frac{32}{1} = \frac{150}{x} \Rightarrow 32x = 150 \Rightarrow x = 4.6875$$

So there are 4 quarts. Separate those off and multiply 0.6875×32 to get the number of liquid ounces:

$$32 \times 0.6875 = 22 \text{ liquid ounces}$$

We know that 4 quarts equals 1 gallon. So 150 liquid ounces equals 1 gallon 22 liquid ounces. Alternatively, we could have done the long division (dividing 32 into 150):

$$32\overline{)150} \\ \quad \underline{128} \\ \quad\; 22 \quad\; \overset{4}{}$$

So 150 liquid ounces equals 4 quarts, 22 liquid ounces or 1 gallon 22 liquid ounces.

Unit 2: Adding and Subtracting Mixed Units

In many business situations, you will need to add and subtract mixed unites. These are not difficult to do. Just be careful.

   Add 5 ft. 3 in. + 6 ft. 7 in. + 4 ft. 5 in.

SOLUTION

First, add the inches: 3 + 7 + 5 = 15 in. That equals 1 ft., 3 in. Now add the feet: 5 + 6 + 4 + 1 = 16 ft.

So the answer is 16 ft. 3 in. We could change it to yards. There are 3 feet in a yard, so the answer is 5 yd. 1ft., 3 in.

   Add 7 lb. 13 oz. + 8 lb. 7 oz. + 5 lb. 11 oz.

SOLUTION

First add the ounces: 13 + 7 + 11 = 31 oz. That equals 1 lb. 15 oz. Now add the pounds: 7 + 8 + 5 + 1 = 21 lb.

So the answer is 21 lb. 15 oz.

   Subtract 20 gross 7 doz. – 8 gross 10 doz.

SOLUTION

It is easiest to set this up like a traditional subtraction problem. We need to borrow a gross from the 20 gross. When the borrowed gross comes over to the dozen column, it equals 12 dozen. So the 7 dozen becomes 19 dozen:

$$
\begin{array}{rr}
\overset{19}{\cancel{20}}\text{g} & \overset{19}{\cancel{7}}\text{d} \\
-8\text{g} & 10\text{d} \\
\hline
11\text{g} & 9\text{d}
\end{array}
$$

So the answer is 11 gross 9 dozen.

EXAMPLE
4

Subtract 19 gal. 2 qt. 1 pt. – 6 gal. 3 qt. 1pt.

SOLUTION

Again, it is easiest to set this up like a traditional subtraction problem. We need to borrow a gallon from the 19 gallons. When the borrowed gallon comes over to the quart column, it equals 4 quarts. So the 2 quarts becomes 6 quarts:

$$
\begin{array}{r r r}
\overset{18}{\cancel{19}}\text{g} & \overset{6}{\cancel{2}}\text{qt} & 1\text{p} \\
-6\text{g} & 3\text{qt} & 1\text{p} \\
\hline
12\text{g} & 3\text{qt} & 0\text{p}
\end{array}
$$

So the answer is 12 gal. 3 qt. 0 pt.

EXERCISES

Exercise A   Find the sums, and simplify your answers.

1. 5 ft. 7 in.
    11 ft. 11 in.
    | 7 ft. 8 in.

2. 5 gross 8 doz.
    7 gross 9 doz.
    +12 gross 10 doz.

3. 8 gal. 3 qt.
    6 gal. 2 qt.
    +12 gal. 3 qt.

4. 12 gal. 2 qt. 1 pt.
    8 gal. 2 qt. 1 pt.
    + 9 gal. 19 qt. 1 pt.

5. 18 gal. 3 qt.
    7 gal. 3 qt.
    +15 gal. 2 qt.

Subtract, then check your answers.

6. 9 ft. 11 in.
    – 5 ft. 6 in.

7. 15 gross 9 doz.
    – 8 gross 11 doz.

8. 15 yd. 2 ft. 10 in.
    – 6 yd. 2 ft. 11 in.

9. 8 gross 5 doz. 9 units
    – 5 gross 8 doz. 11 units

10. 8 gross   9 units
    – 3 gross 5 doz. 10 units

Word Problems

Exercise B   Solve the following problems:

11. From a 50-pound bag of rice, a grocer sold the following weights: 6 pounds 10 ounces, 4 pounds 11 ounces, and 8 pounds 14 ounces. How many pounds of rice are left in the sack?

12. Ribbon sells for $2.75 per foot. What is the total cost of the following lengths of ribbon: 2 yards 1 foot, 3 yards 18 inches, 4 yards 30 inches?

13. Pedro bought the following weights of cold cuts: 1 pound 10 ounces of ham, 1 pound 6 ounces of salami, and 1 pound 9 ounces of roast beef. What was the total weight of all the cold cuts?

Unit 3: Multiplying and Dividing Mixed Units

Many situations in business call for you to multiply and divide mixed units of measure. For example, a bolt of cloth (120 feet long) arrives at a factory. How many dresses can be made from that one bolt if one dress takes 2 yd. 1 ft. of cloth to make?, If a suit takes 5 yd. 8 in. of cloth to make, how much fabric is needed to make a dozen suits?

Let's do the multiplication problem just described. We want to make a dozen suits, and each suit requires 5 yd. 8 in. of fabric. How much cloth will we need?

SOLUTION

To solve this, multiply $5 \times 12 = 60$ yd. and $8 \times 12 = 96$ in.
We need to convert 96 inches into yards. There are 36 inches in a yard.

$$2 \times 36 \text{ in.} = 72 \text{ in.} = 2 \text{ yd.}$$

We can take 2 yards out of the 96 inches and get $96 - 72 = 24$ in. left. Since there are 12 inches in a foot, we have 2 feet left.
Add the yards: $20 + 2 = 22$ yd. Add the feet to the yards: 22yd. 2ft. We will need 22 yd. 2 ft to make a dozen suits.

Steve is a manager planning for a breakfast meeting and wants bacon on the menu for the participants. Each person, on average, will eat about 4 oz. of bacon. There are 35 people coming. How many pounds of bacon will the manager need to buy?

SOLUTION

To solve this, we multiply $35 \times 4 = 140$ oz. We need convert this to pounds: $140/16 = 8.75$lb. Converting the 0.75 lb. back to ounces is $0.75 \times 16 = 12$ oz. This means that the manager will need to order 8 lb. 12 oz. of bacon to provide for the meeting.

Alternatively, we could divide:

$$
\begin{array}{r}
8 \\
16\overline{)140} \\
\underline{128} \\
12
\end{array}
$$

This gives us 8 lb. with 12 oz. remainder.

Ginger, a manager at a dress factory, gets a bolt of cloth 120 feet long. She uses 2 yd. 1 ft. of cloth to make one dress. How many dresses can she make with the bolt of cloth? How much cloth will be left over?

SOLUTION

First convert the amount of cloth needed to make one dress into feet: 2 yd. = 6 ft. and 1 more foot is 7 ft. Then divide 120 ft. by 7 ft.:

$$
\begin{array}{r}
17 \\
7\overline{)120} \\
\underline{7} \\
50 \\
\underline{49} \\
1
\end{array}
$$

This shows that Ginger will be able to make 17 dresses, and there will be 1 ft. of cloth left over.

Pat is a bar owner. She receives a keg of beer, which contains 16 gal. How many 12 oz. glasses can she draw (pour) out of the keg?

SOLUTION

First, note that there are 128 oz. in a gallon. Then multiply 16 by 128 to get the number of ounces in the barrel: $16 \times 128 = 2048$. Now divide 2048 by 12 to get the number of glasses Pat can get out of the barrel:

$$
\begin{array}{r}
170 \\
12\overline{)2048} \\
\underline{12} \\
84 \\
\underline{84} \\
08
\end{array}
$$

The answer is 170 12 oz. glasses of beer will come out of a 16-gallon keg, with 8 oz. left over.

AREA PROBLEMS

Area is the measure of a two-dimensional space, such as a floor or a wall. To find the area, multiply the two dimensions of the space. For instance, suppose a floor is 30 ft. long and 20 ft. wide. Then the area of the floor is $30 \times 20 = 600$ square ft. Area is always named as a square quantity. In the above example, another way of writing this is 600 ft.^{2} Here are some examples of square measure:

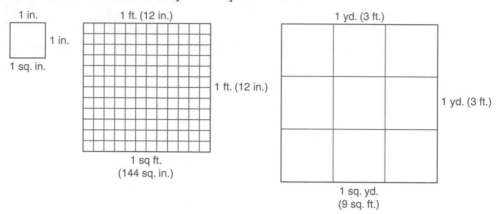

To find the area, measure the length and the width. Then multiply them together. The formula is:

$$A = L \times W$$

EXAMPLE 5

What is the area of a warehouse floor that is 30 yd. long and 75 ft. wide?

SOLUTION

First, we must make both of the quantities the same units. 30 yd. is fine the way it is. Since there are 3 ft. to a yd., 75 ft. converted to yards is $75/3 = 25$ yd. Now we can multiply:

$$30 \times 25 = 750 \text{ square yd.} = 750 \text{ yd.}^2$$

We can write it either way, but the second way is the most common.
We can change the yd.^{2} to ft.^{2} by multiplying by the conversion factor: 9 ft.^{2}

$$750 \times 9 = 6750 \text{ ft.}^2$$

EXAMPLE 6

A wall is 10 ft. high and 7 yd. long. What is the area of the wall in square feet?

SOLUTION

This is a bit tricky because we do not have an even number of yards.

$$10/3 = 3.333 \text{ yd.}$$
$$3.333 \times 7 = 23.33 \text{ yd.}^2$$

To convert to this to square feet, we must multiply by 9:

$$23.33 \times 9 = 209.97$$

209.97 ft.² rounds to 210 ft.²
Alternatively, we could change all the units to feet:

$$7 \text{ yd.} \times 3 \text{ ft.} = 21 \text{ ft.}$$
$$21 \times 10 = 210 \text{ ft.}^2$$

This solution agrees with our other solution above.

 EXAMPLE 7 Tom, an entrepreneur, is using a warehouse to set up his business. The warehouse is 250 ft. wide and 300 ft. long. Tom does not need all of that room, however. He needs only 16,000 ft.² to set up his business. How many square feet is the warehouse, and where should Tom put a partition up to contain his 16,000 ft.²?

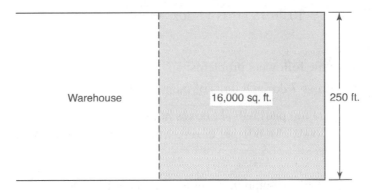

SOLUTION

First, calculate the area of the warehouse:

$$300 \times 250 = 75,000 \text{ ft.}^2$$

Next, calculate how long the warehouse needs be to get 16,000 ft.² We need to use 250 ft. as width, because that is the width of the warehouse. We know the area: 16,000 ft.² We just need to calculate the length. Use division:

$$(250)(x) = 16,000 \Rightarrow x = 16,000/250 \Rightarrow x = 64$$

Tom will need to measure out 64 ft. to get an area of 16,000 ft.² in the warehouse.

EXERCISES

Exercise A   Multiply and simplify.

1. 3 ft. 9 in.
   × 5

2. 4 yd. 2 ft. 7 in.
   × 8

3. 6 gal. 3 qt. 1 pt.
   × 8

4. 8 lb. 15 oz.
   × 6

5. 2 yd. 8 in.
   × 12

Divide and simplify. Round off each answer to the nearest whole smaller unit.

6. 5)9 ft. 10 in.

7. 4)12 lb. 8 oz.

8. 12)30 gal. 3 qt.

9. 7)14 lb. 15 oz.

10. 6)15 gross 11 doz. 8 units

Word Problems

Exercise B   Solve the following problems:

11. A box holds 3 gross 7 dozen 8 units of memo pads.

    (a) How many memo pads will six boxes hold?
    (b) If the memo pads cost 60¢ per dozen, what is the cost of six boxes of memo pads?

12. A typewriter packed for shipping weighs 13 pounds 14 ounces.

    (a) What is the total weight of six typewriters?
    (b) If the cost of shipping each typewriter is $9.95, what is the total cost of the shipment?

13. A drum of liquid adhesive containing 20 gallons 3 quarts is to be poured into eight containers. How much would each container hold? (Round to the nearest quart.)

Unit 4: The Metric System

The metric system is named for the fundamental unit of linear measure, the meter. Most countries in the world use the metric system. Increasingly, the United States is adopting the metric system. Great Britain, the country most responsible for the proliferation of the English system of weights and measures, is using the metric system more frequently.

The metric system is a decimal system. There are three fundamental units of measure, and then a prefix is placed onto that unit of measure to indicate another measure. For instance, the meter, the linear measure, is a little over 39 in. long (or about 1 yd. 3 in.) The prefix "kilo" is added to the meter to get the kilometer. This is the unit of linear measure that corresponds to the mile in the English system. The following tables show different aspects of the metric system.

PREFIXES USED IN THE METRIC SYSTEM

Prefix	Definition
Deca	Ten (10)
Hecto	One hundred (100)
Kilo	One thousand (1000)
Deci	Tenth (0.1)
Centi	Hundredth (0.01)
Milli	Thousandth (0.001)

THREE FUNDAMENTAL UNITS OF MEASURE IN THE METRIC SYSTEM

	Unit	U. S. Equivalent
Length	Meter (m)	39.37 inches
Volume	Liter (L)	1.06 quarts
Weight	Gram (g)	0.04 ounce

TABLE 2
METRIC UNITS OF MEASURE

Units of Length Meter (m)	
Larger Units	**Smaller Units**
10 meters = 1 decameter (dam)	$0.1 \left(\dfrac{1}{10} \right)$ meter = 1 decimeter (dm)
100 meters = 1 hectometer (hm)	$0.01 \left(\dfrac{1}{100} \right)$ meter = 1 centimeter (cm)
1,000 meters = 1 kilometer (km)	$0.001 \left(\dfrac{1}{1,000} \right)$ meter = 1 millimeter (mm)

(Continued)

Units of Capacity (Liquid)
Liter (L)

10 liters = 1 decaliter (daL)

$0.1 \left(\dfrac{1}{10} \right)$ liter = 1 deciliter (dL)

100 liters = 1 hectoliter (hL)

$0.01 \left(\dfrac{1}{100} \right)$ liter = 1 centiliter (cL)

1,000 liters = 1 kiloliter (kL)

$0.001 \left(\dfrac{1}{1,000} \right)$ liter = 1 milliliter (mL)

Units of Weight
Gram (g)

10 grams = 1 decagram (dag)

$0.0 \left(\dfrac{1}{10} \right)$ gram = 1 decigram (dg)

100 grams = 1 hectogram (hg)

$0.01 \left(\dfrac{1}{100} \right)$ gram = 1 centigram (cg)

1,000 grams = 1 kilogram (kg)

$0.001 \left(\dfrac{1}{100} \right)$ gram = 1 milligram (mg)

CHANGING UNITS IN THE METRIC SYSTEM

Changing units in the metric system is easy. Every conversion factor is a power of 10 (10, 100, 0.0, 0.0001, and so on).

Change 25,000 grams to kilograms.

SOLUTION

1 kilogram equals 1,000 grams. To change grams to kilograms, we must divide by 1,000:

$$\frac{25,000}{1,000} = 25$$

25,000 grams equals 25 kilograms.

Alternatively, we could move the decimal place three places to the left to indicate dividing by 1,000:

$$25{,}000\text{g} \Rightarrow 25\text{kg}$$

Again, we see that 25,000 grams equals 25 kilograms.

EXAMPLE 2   Change 3 liters to deciliters.

SOLUTION

1 liter equals 10 deciliters. To change liters to deciliters, we must multiply by 10:

$$3 \times 10 = 30$$

3 liters equals 30 deciliters.

Alternatively, we could move the decimal place one place to the right to indicate multiplying by 10:

30

Again, we see that 3 liters equals 30 deciliters.

EXERCISES

Exercise A   By moving the decimal point, you should be able to perform most of the multiplications and divisions without written calculations. Change:

1. 240 cm to meters
2. 0.53 m to millimeters
3. 573 mL to liters

4. 4.15 cL to liters
5. 53 cg to milligrams
6. 4,375 mm to meters

7. 0.6 g to milligrams
8. 0.08 km to meters
9. $23\frac{1}{2}$ g to centigrams

10. $5\frac{1}{4}$ L to centiliters
11. 235 L to centiliters
12. 432 cg to kilograms

13. 0.05 kL to milliliters
14. 0.4 km to meters
15. 635 cm to millimeters

Word Problems
Exercise B   Solve the following problems:

16. A vial holds 55 milliliters. How many liters will 48 vials contain?

17. How many 50-mL vials can be filled with $2\frac{1}{2}$ liters of liquid?

18. A piece of tubing measures 1.5 m. How many pieces of 50-cm length can be cut?

Unit 5: Metric–English Conversions

The United States is converting to the metric system. However, conversion takes time and money. Factories must retool their machines, their tools, and their computers. Cost estimates are in the billions. In the meantime, conversions must be made from English to metric and vice versa. Tables 3 and 4 give the English–metric and metric–English conversions.

TABLE 3 ENGLISH TO METRIC	TABLE 4 METRIC TO ENGLISH
Metric Equivalents of English Measures	*English Equivalents of Metric Measures*
1 in. = 25.4 mm = 2.54 cm	1 m = 39.37 in = 3.28 ft. = 1.09 yd.
1 ft. = 0.30 m	1 cm = 0.39 in.
1 yd. = 0.91 m	1 mm = 0.039 in.
1 mi. = 1.61 km	1 km = 0.62 mi.
1 qt. (liquid) = 0.95 L	1 L = 1.06 qt. (liquid)
1 qt. (dry) = 1.1 L	1 L = 0.91 qt. (dry)
1 oz. = 28.35 g	1 g = 0.04 oz.
1 lb. = 0.45 kg	1 kg = 2.2 lb.
1 s.t. = 0.91 metric ton	1 l.t. = 2,240 lb

EXAMPLE 1

Convert 10 meters to feet.

SOLUTION

1 m equals 3.28 ft.

$$10 \times 3.28 = 32.8$$

10 meters equals 32.8 feet.

EXAMPLE 2

How many pounds are in 5 kilograms?

SOLUTION

There are 2.2 lb. in 1 kg.

$$5 \times 2.2 = 11$$

5 kilograms equals 11 pounds.

Convert 25 L to gal.

SOLUTION

We need to convert to quarts first and then to gallons. There are 0.91 qt. in 1 L.

$$25 \times 0.91 = 22.75$$

So there are 22.75 qt. in 25 L. There are 4 quarts in a gallon

$$20/4 = 5 \text{ gal.}$$

2.75 qt. are left over. 25 L equals 5 gal. 2.75 qt.

How many grams are in 3 pounds?

SOLUTION

We will need to convert the pounds to ounces first and then to grams. There are 16 oz. in a pound.

$$16 \times 3 = 48 \text{ oz.}$$

There are 28.35 g in an oz.

$$48 \times 28.35 = 1{,}360.8 \text{ g}$$

3 lb. equals 1,360.8 g.

How many feet are in 1,610 meters?

SOLUTION

1 meter equals 3.28 ft.

$$1{,}610 \times 3.28 = 5{,}280.8 \text{ ft.}$$

1,610 m is 5,280.8 ft. Does this number of feet look familiar? It is the number of feet in a mile (5,280 feet equals 1 mile). So 1,610 meters is about 1 mile.

EXERCISES

Exercise A   Convert as indicated, rounding each answer to the nearest whole number.

1. 20 m to feet

2. 130 km to miles

3. 42 oz. to grams

4. 115 km to miles

5. 15 dry qt. to liters

6. $3\frac{1}{2}$ kg to pounds

7. 2,300 lb. to metric tons

8. $3\frac{1}{4}$ in. to millimeters

9. 2.3 ft. to decimeters

10. 86 oz. to kilograms

11. 32.8 cm to inches (nearest *tenth*)

12. $\frac{3}{4}$ kg to pounds (nearest *hundredth*)

Word Problems

Exercise B   Solve the following problems:

13. What is the width in inches of 35-mm film? (nearest *hundredth*)

14. How many liters are there in 5 quarts of milk? (nearest *tenth*)

15. A room measures 3.6 m by 6.5 m. What are the room measurements in feet? (nearest *tenth*)

UNIT 6: REVIEW OF CHAPTER 4

Terms:
- Mixed units
- Distance—linear measures
- Area—square measures
- Meter, liter, gram
- Deka- (10), hecto- (100), kilo- (1,000)
- Deci $\left(\frac{1}{10}\right)$, centi- $\left(\frac{1}{100}\right)$, milli- $\left(\frac{1}{1,000}\right)$

Hints:
- When changing *smaller* units to *larger* units, *divide*.
- When changing *larger* units to *smaller* units, *multiply*.
- When possible, change mixed units to mixed fractions of the larger unit.

Add or subtract as indicated and simplify your answer.

1.   4 yd. 2 ft. 11 in.
      12 yd. 1 ft.   9 in.
   +13 yd. 2 ft. 10 in.

2.   14 gross 6 doz.   8 units
    − 4 gross 9 doz. 10 units

Multiply or divide as indicated. Simplify and round to the nearest whole smaller unit.

3. 8 gal. 2 qt. 1 pt.
   ×   13

4. 14)$\overline{34 \text{ gross 9 doz. 6 units}}$

Change the following metric units without using written calculations:

5. 0.8 g to mg

6. $8\frac{1}{4}$ L to cL

7. 0.73 km to m

8. 115 cL to mL

Change the following units and round each answer to the nearest whole unit:

9. 53 oz. to g   10. 42.7 km to mil.   11. $1\frac{1}{8}$ in. to mm

Solve these area problems:

12. An office measures 12 ft. 6 in. by 16 ft. 9 in. How many square yards of carpeting are needed to cover the floor?

13. A large office is to be subdivided into four equal work stations. If the office measures $9\frac{1}{2}$ meters by $14\frac{1}{4}$ meters, what will be the area of each work station?

Banking and Investments

Saving money is convenient and safe in a bank, a savings and loan association, or a credit union. Money is deposited, and the Federal Deposit Insurance Corporation (FDIC) guarantees deposits. The bank (or other institution) gives interest (money) on the balance in the account. The bank can do this because it will lend the money in turn to the government, to the commercial sector, or for a personal loan. Thus the bank makes money for itself.

Unit 1: Savings Accounts

There are two main types of savings accounts: **time deposit accounts** and **withdrawal accounts**. Most people have the latter. These days, they are associated with bank cards. The withdrawal account allows the depositor to withdraw money on demand with no real penalty. Interest is paid on a withdrawal account (sometimes called a day-of-deposit account), but it is not a high rate of interest. Time deposit accounts (often known as deposit accounts) generally offer a higher rate of interest, but the money is locked into the account. That is, the depositor must wait a set amount of time before the money is released (usually a month or two). If the depositor withdraws the money early, a penalty is imposed. The penalty is usually equal to the interest that would have been earned. Another financial instrument, the certificate of deposit (CD), is also common. The depositor agrees to leave the CD in the bank for a longer period of time (6 months is typical), and the depositor gets a higher rate of interest as a result. When the 6-month period is over, the CD is said to mature. The depositor gets his/her money plus the interest that was promised. As with deposit accounts, a penalty is incurred if the money is withdrawn early, sometimes equal to the interest, sometimes more. You should remember that not all bankers are familiar with all these different names. Check with your banker to find out what he/she calls each account.

CALCULATING INTEREST

Interest, as mentioned before, is money given to a depositor who leaves money in an account of some kind (savings, deposit, and so on). Interest is usually calculated as a percentage of the amount of money in the account. To use a very simple example, suppose a depositor leaves $100 in an account for one year that offers a 2% interest rate. At then end of the year, the depositor will earn:

$$100 \times 0.02 \times 1 = \$2$$

The depositor will earn 2 dollars on 100 dollars after one year. It does not seem like a lot, but interest builds up over time. The full equation for calculating interest is:

$$\underset{\text{Principal}}{P} \times \underset{\text{Rate}}{R} \times \underset{\text{Time}}{T} = \underset{\text{Interest}}{I}$$

- *Principal* is the money originally deposited ($100 in the example).
- *Rate* is the annual interest rate (2% in the example).
- Time is the amount of time the principal stays in the account (one year in the example).
- *Interest* is the amount of money earned ($2 in the example).

Because the interest rate is an annual rate, the period of time for calculating interest must be in years or in fractions of years.

Two kinds of interest are used in banking: **simple interest** and **compound interest**. Simple interest was shown in the example. It is calculated on a fixed amount of money. In the example, every year $100 will earn $2 for the depositor per year. With compound interest, the amount of interest ($2) is added to the principal. In the second year, the interest is calculated on $102. It is not hard to see that with compound interest, the principal in the account grows more rapidly than with simple interest. As another example, consider the following chart:

COMPOUND INTEREST				SIMPLE INTEREST	
Principal	Today's (Day 1) Interest		Principal	Cumulative Interest	Today's (Day 1) Interest
$1,000	$0.15	Day 1	$1,000		$0.15
1,004.53	0.15	Day 30	1,000	$ 4.37	0.15
1,013.65	0.15	Day 90	1,000	13.41	0.15
1,056.54	0.16	Day 365	1,000	54.85	0.15
$1,056.54			Total principal and interest	$1,055	
5.654%			Effective annual yield	5.5%	

Note: **Effective annual yield** means that the annual interest rate compounded daily is equal to a higher simple interest rate. In this example, a 5.5% rate of interest compounded daily is equal to (has an effective annual yield of) 5.654%. Note also that, if you were to withdraw your interest from a compound interest account, your effective annual yield would be the same as under simple interest.

For the following examples, we will be calculating simple interest. That is, we will not be adding the interest into the principal to calculate subsequent interest in these examples.

Betsy put $850 into a day-of-deposit account at 5% interest. How much money will she have at the end of the year?

SOLUTION

$$P \times R \times T = I$$
$$\$850 \times 0.05 \times 1 = I \Rightarrow I = \$42.50$$

New balance: $850 + $42.50 = $892.50.

Note that when we calculate interest for one year, the "time" is 1. Remember that anything multiplied by 1 is that same number. For this reason, if the calculation is for one year, the "time" term is left out of the equation.

Alternatively, we could use the percent equation:
What number is 5% of 850?

$$\frac{5}{100} = \frac{x}{850} \Rightarrow 100x = (5)(850) \Rightarrow x - 4{,}250/100 \Rightarrow x - 42.50$$

New balance: $850 + $42.50 = $892.50.

Marcos put $1,275 into a withdrawal account at 4.5% interest. He took it out after 6 months. How much will he have at the end of the half year?

SOLUTION

$$P \times R \times T = I$$
$$\$1{,}275 \times 0.045 \times 0.5 = I \Rightarrow I = 28.6875$$

By rounding off, we have $I = \$28.69$.
New balance: $1,275 + 28.69 = $1,303.69.

Alternatively, we could use the percent equation.
What number is 4.5% of 1,275?

$$\frac{4.5}{100} = \frac{x}{1{,}275} \Rightarrow 100x = (4.5)(1{,}275) \Rightarrow x = 5{,}737.5/100 \Rightarrow x = 57.375$$

This amount is for one year, but Marcos has the principal in for half a year. So we must divide by 2:

$$57.375/2 = 28.6875$$

Rounding off gives 28.69.
New balance: $1,275 + $28.69 = $1,303.69

Mahmoud has $2,835 in a savings account at 4% interest. After 60 days, how much interest has he accrued?

Note: Banks consider a year to be 360 days instead of 365.

SOLUTION

$$P \times R \times T = I$$
$$\$2{,}835 \times 0.04 \times 60/360 = I \Rightarrow I = 18.9$$

Interest = $18.90   New balance: $2,835 + $18.90 = $2,853.90

Alternatively, we could use the percent equation. What number is 4% of 2,835?

$$\frac{4}{100} = \frac{x}{2{,}835} \Rightarrow 100x = (4)(2{,}835) \Rightarrow x = 11{,}340/100 \Rightarrow x = 113.40$$

This amount is for one year. However, Mahmoud has the principal in for 60 days, which is 60/360 = 1/6. We divide the interest amount by 6:

$$113.4/6 = 18.9$$

New balance: $2,835 + $18.90 = $2,853.90

Anastasia put $2,537 into a withdrawal account. She left the money in for half a year and received $63.43 in interest. What interest rate was she earning?

SOLUTION

$$P \times R \times T = I$$
$$\$2{,}537 \times R \times 180/360 = \$63.43$$

Solving for R gives:

$$R = 63.43/2{,}537 \times 360/180 \Rightarrow R = 0.0500039$$

This rounds to 0.05. So the interest rate was 5%.

Alternatively, we could have used the percent equation:
What percent of 2,537 is 63.43?

$$\frac{x}{100} = \frac{63.43}{2{,}537} \Rightarrow 2{,}537x = (100)(63.43) \Rightarrow x = 6{,}343/2{,}537 \Rightarrow x = 2.500197$$

This rounds to 2.5. However, Anastasia had the money in only half a year, so we must double this to find the interest rate:

$$2.5 \times 2 = 5$$

This account pays 5% interest.

EXAMPLE 5 Dr. Bianchi put $20,000 in a 6-month certificate of deposit. The CD pays an interest rate of 13.85% per annum (year). The effective annual yield is 14.5%.

a. How much interest will he earn if he withdraws his money every month?

SOLUTION

Since Dr. Bianchi is withdrawing the interest each month, the base interest rate of 13.85% is used:

$$P \times R \times T = I$$
$$20,000 \times .1385 \times 0.5 = \$1,385$$

Dr. Bianchi will earn $1,385 in interest.

b. If he leaves the money in the account for the full 6 months, how much interest will he earn?

SOLUTION

Since Dr. Bianchi leaves the money in for the full amount of time on the CD, he will earn interest at the effective annual yield of 14.5%:

$$P \times R \times T = I$$
$$20,000 \times .145 \times 0.5 = \$1,450$$

Dr. Bianchi will earn $1,450 at the maturity of this CD.

COMPUTER ACCESS TO SAVINGS ACCOUNTS

Banks have provisions for you to view your account information online these days. You should consult your bank about how to access the online features they extend.

EXERCISES

Exercise A   Each of the following accounts had no withdrawals or deposits for the period shown. Find the amount of interest and the new balance for each account.

	Amount	Interest Rate	Period of Time	Amount of Interest		New Balance	
1.	$ 8,400	5.5%	2 years	_____	___	_____	___
2.	11,000	$5\frac{1}{2}$	1 year	_____	___	_____	___
3.	5,800	$5\frac{1}{4}$	2 years	_____	___	_____	___
4.	10,000	$8\frac{1}{2}$	4 years	_____	___	_____	___
5.	11,765	$9\frac{1}{4}$	2 years	_____	___	_____	___

	Amount	Interest Rate	Period of Time	Amount of Interest		New Balance	
6.	$15,000	6%	3 months	_____	___	_____	___
7.	8,440	$8\frac{1}{4}$	6 months	_____	___	_____	___
8.	18,750	8.5	8 months	_____	___	_____	___
9.	14,725	11.75	2 months	_____	___	_____	___
10.	24,000	$10\frac{1}{4}$	15 days	_____	___	_____	___

Fill in the blanks in the following table:

	Amount	Interest Rate	Period of Time	Amount of Interest	New Balance	
11.	$5,850	_____	1 year	$ 394.88	_____	___
12.	4,675	_____	1 year	245.44	_____	___
13.	_____	11.63	1 year	1,133.93	_____	___

Find the interest earned for each of the following accounts, using the appropriate interest rate—the annual rate *or* the effective yield.

	Amount	Certificate of Deposit	Annual Rate	Effective Yield	Interest Withdrawn	Interest Earned	
14.	$12,000	30 months	12%	12.94%	Yes	_____	___
15.	4,000	30 months	12.5	13.475	No	_____	___
16.	14,250	6 months	13.95	14.63	Yes	_____	___
17.	8,500	30 months	12.5	13.475	Yes	_____	___
18.	12,800	6 months	12.125	13.89	No	_____	___

Fill in the following table, showing the amount of interest and new balances on a 6-month CD.

	Amount on Deposit		Interest Rate	Amount of Interest		New Balance at End of 6 Months	
19.	$1,465	75	8.752%	_____	___	_____	___
20.	2,528	80	8.025	_____	___	_____	___
21.	4,521	48	10.035	_____	___	_____	___

Word Problems

Exercise B   Solve the following problems:

22. Meghan opened a savings account, paying 6% interest, with $5,300. Six months later she deposited another $1,000. If she does not withdraw interest for the entire year, how much money will she have in her account at the end of a year?

23. Rachel has a savings account of $18,000. If she earned $2,610 in 1 year, what rate of interest did the bank pay her?

24. Sara put $15,000 into a 6-month certificate of deposit, paying an annual rate of 12.65%, with an effective annual yield of 13.7%. After 6 months, she renewed her certificate of deposit, and the new interest rate was 13.25% annual rate, with an effective annual yield of 14.85%. How much money will Sara have at the end of a year if she does not withdraw any interest?

Unit 2: Checking Accounts

A checking account is an account in a bank in which you have immediate use of the funds deposited there. The bank gives you a set of checks and deposit slips with your name and address on them. You use the account to pay bills by mail, pay for memberships in a club, or buy items in stores.

When a checking account is opened, you can have your paychecks deposited directly into the account and have fast access to that money. In addition to checks, most checking accounts will come with a **bank card (debit card)**. It looks much like a credit card. This card can be used instead of writing a check. Merchants will swipe the card in a card reader, and the money will be automatically transferred to the merchant's account without you needing to write a check.

You can deposit checks that have been sent to you. However, a bank hold checks for a specified time. So you might not have access to that money right away. You must be careful not to write checks for money that is not in your checking account. The bank may not pay the company you issued the check to, marking the check you wrote "Insufficient Funds." Additionally, the bank will charge you a penalty for writing the bad check.

Not all checking accounts pay interest, but some do. Some checking accounts charge service fees and per-check fees, and some do not. Some banks charge you for the checks you order, and some do not. There are personal and commercial checking accounts. You should check with your bank and be certain of all the terms and conditions before you open a checking account.

MAKING A DEPOSIT

You can deposit cash and checks into a checking account using a **deposit slip**. You can use preprinted deposit slips or use deposit slips in the bank. Make sure, though, that your own checking account number is on the deposit slip. If it is not, your money will go into another person's account. As an example, suppose you have this cash, coin, and checks to deposit:

- Three $20 bills
- Five $10 bills
- Two $5 bills
- Three $1 bills
- Five quarters
- A paycheck for $92.65
- Three more checks in the amounts of $35.85, $63.70, and $25.80

Lets quickly add up the cash:

$$60 + 50 + 10 + 3 + 1.25 = \$124.25$$

This is how your deposit slip would look:

	DEPOSIT TICKET	
	DOLLARS	CENTS
CASH	124	25
CHECKS LIST SINGLY	92	65
	35	85
	63	70
	25	80
TOTAL	342	25

THE CITY TRUST COMPANY
Staten Island, N.Y.

JOHN KLUG

DATE *July 5,* 20==

⑆585072 10⑈

Next, you need to endorse each of the four checks you wish to deposit. To endorse them, turn each check over. Make sure the top of the back corresponds to the left side of the front. Sign it on the top of the reverse side:

PAUL M. MURDOCH   NO. 311

Mar 8 20==

PAY TO THE ORDER OF *John Klug* $ *92.65*

Ninety-Two and 65/100 ———— DOLLARS

ISLAND BANK
Staten Island, N.Y.

FOR *Car Repairs*   *Paul M. Murdoch*

⑆022558⑆ 500⑈

John Klug

Finally, you would bring all the cash and the checks to the bank teller, and get a deposit receipt. As mentioned before, the funds may not be immediately available. Sometimes banks hold a check for 2–3 days, sometimes longer. This action is sometimes referred to as a **credit transaction**.

WRITING A CHECK

When you pay bills or pay a merchant with a check, you use the checks supplied to you by the bank. In the book of checks, there is usually a register for you to record all the particulars of the check you wrote. For example, suppose you need to give a rent check to the Ace Realty Co. for $175.00. Here is how the check would look:

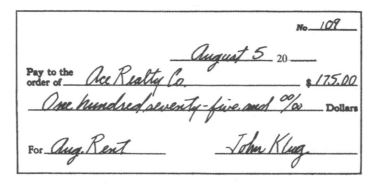

Notice that the check has these parts:

1. *Check number*: The number of the check up at the top. Often the check will be numbered by the bank.

2. *Check date*: The date that you actually wrote the check. Be careful to enter the right date. A check that is six months after the date on the check is considered a "stale check," and a bank may not cash it.

3. *Pay to the order of*: Who the check is made out to (the payee). This authorizes your bank to pay the Ace Realty Co. out of your checking account. Usually this is a person or an organization.

4. *Amount*: Right next to the payee line, you must put the amount of the check in numbers.

5. *Written amount*: Right under the payee line, you must write out the amount of money in *words*. This is important in banking. Writing out the amount in numbers and in words makes the check hard to be changed once you have written it. Always write out the words of the dollar amount, and then XX/100 for the cents because all checks say "Dollars" at the end. (Remember to put the actual number of cents over 100 and not XX.)

6. *For*: This reminds you of what you wrote the check for. Some checks will call this "*Note*." You may not see this check again for 30-40 days. Filling in the section will be very helpful to you later. It will help you remember your purchases and what you wrote the check for.

7. *Signature*: When you opened the checking account, the bank asked you for your signature. They will check your signature on this check against the signature they have on record for you.

In the book of checks, the register is usually separate from the checks (although sometimes there is a copy leaf behind the check that records all the information you wrote on the check). Make sure you record your checks in the register and subtract the latest check from your previous balance to get the "balance carried forward."

BALANCING A CHECKBOOK

Usually once a month, you receive a **bank statement**, which informs you of all the transactions you made in your checking account. It tells you the beginning and ending balance (money) in your account. It also tells you all the checks you wrote against the balance (known as **debits**), all the deposits you made in the month (known as **credits**), and any fees you were charged (a bank fee, an overdraft fee, and so on). These fees are also known as **debits**).

Very often, the balance you think is in your account does not agree with the actual balance you see when the statement comes at the end of the month. There are a lot of reasons for this, such as charges you had not known were there, checks not being deposited by the party you issued them to, arithmetic errors, and so on.

To balance a checkbook, follow these steps:

1. Check the previous balance (from the beginning of the month).

2. Arrange all the checks written in order, by date.

3. Verify all dates of other transactions (deposits, fees, penalties).

4. Go down the bank statement, verifying each amount (either debit or credit).

5. Make a note of any fees or penalties you did not know about or do not understand.

6. Check the ending balance to see if you agree with the bank's statement.

7. If there is a difference, there is an error in either the bank statement or in the checkbook.

8. Subtract the difference between the checkbook and the bank statement.

9. To find the error, look at the difference you got in step 8:

   • Carefully look through the bank statement. Sometimes the error is a fee or a charge you overlooked.
   • The discrepancy could be a check you missed or one that was not returned to you.
   • Check your addition and subtraction.

COMPUTER STATEMENTS

These days, you can do your banking online and get a statement updated daily. You do not have to wait until the end of the month or until the mail comes with the statement. Check with your bank to find out how to access their online banking service.

EXAMPLE 1   Suzanne's bank statement shows a balance of $483.55, but her checkbook shows a balance of $241.70. After checking her canceled checks against the checkbook register, she found that three checks had not been charged against her account. The three checks were in the amounts of $48.75, $35.50, and $65.10. In addition, there was a deposit of $95.50 Suzanne forgot to include in her register. Also, the bank charged a $3 service fee. Knowing all that, Suzanne was ready to make up a *reconciliation statement*:

Checkbook Balance		Bank Statement Balance	
	$241.70		$483.55
Add:		*Add:*	
Deposit omitted in register	95.50		
	$337.20		
Deduct:		*Deduct:*	
Service charge	3.00	Checks outstanding	149.35
Adjusted balance	$334.20	Adjusted balance	$334.20

EXERCISES

Exercise A   Using the forms provided, fill out a deposit slip for each of the following:

1.
Please endorse all checks and list below singly			
DATE	, 20	DOLLARS	CENTS
	BILLS		
	COINS		
	CHECKS		
	TOTAL		

2.
Please endorse all checks and list below singly			
DATE	, 20	DOLLARS	CENTS
	BILLS		
	COINS		
	CHECKS		
	TOTAL		

3.

Please endorse all checks and list below singly			
DATE	, 20	DOLLARS	CENTS
	BILLS		
	COINS		
	CHECKS		
	TOTAL		

1. Bills: 3 \$20's, 2 \$10's, 4 \$5's, 7 \$1's
   Coins: 6 quarters, 7 dimes, 8 nickels, 17 pennies
   Checks: \$85.70, \$24.50, \$17.05

2. Bills: 5 \$20's, 5 \$10's, 6 \$5's, 11 \$1's
   Checks: \$98.75, \$63.15, \$35.38, \$18

3. Bills: 3 \$20's, 4 \$10's, 5 \$5's, 12 \$1's
   Checks: \$73.30, \$84.75

Using the blank forms provided, prepare checks according to the information in Problems 4 through 6. The initial balance brought forward is \$219.85; find each succeeding balance carried forward.

4.

5.

6.

4. Check #110:   To Consolidated Utility Co. for July electric bill; amount of check, $48.75; date of check, August 5.

5. Check #112:   Deposit, $315; amount of check, $215; to Ace Realty for August rent; date, August 8.

6. Check #114:   Amount of check, $55; to Lang's Department Store for charge account payment; date, August 11.

Exercise B   Prepare a reconciliation statement for each of the following accounts, using the reconciliation statement on page 150 as a guide. What is the corrected balance for each?

	Checkbook Balance	Bank Statement Balance	Outstanding Checks	Service Charge	Other Adjustments	Corrected Balance
7.	$313.64	$365.39	$23.50 20.75 75.65	$3.15	$65 deposit recorded twice	_____
8.	$146.18	$324.26	$27.13 23.25 30.70	$3.00	$100 deposit omitted from check stub	_____
9.	$281.90	$289.90	$ 63.25 101.50 25.25	$2.00	$180 deposit not on bank statement	_____
10.	$259.25	$369.10	$128.85 64.30 20.70	$4.00	$100 deposit recorded twice on check stubs	_____
11.	$662.40	$557.20	$ 75.60 109.62 71.83	$3.50	$258.75 deposit not on bank statement. $100 check omitted on checkbook stub	_____

Word Problems

Exercise C Solve the following problems:

12. Henry's checking account showed a balance brought forward of $243.65. He wrote the following checks: $175.50; $68.78; $47.65; and $109.15. If he deposited $87.50 in cash plus his paycheck of $265.80, what is his new balance brought forward?

13. Tony's checkbook balance was $321.75 but he had neglected to record a check for $225. If he then writes checks for $132.50, $78.95, and $63.75, and deposits $235.65, what is his current checkbook balance?

14. Raphael had $215.72 in his checking account and wrote checks for $63.75, $49.75, and $57.50. Find his new balance if he deposited $186.90 in his account.

Prepare a reconciliation statement for the following account:

15. Marcia received her bank statement for July, showing a balance of $429.71. Her check stubs showed a balance of $45.30. In reviewing her checkbook and the canceled checks, she found three checks outstanding for $125.63, $65.10, and $23.18. She had also omitted from her checkbook register a deposit of $175. The bank service charge was $4.50. What is her corrected balance?

Unit 3: Investments–Stocks

One of the fundamental supports of our economy is the stock market. Corporations that want to raise money to build a factory or to expand into new areas of the country or the world issue stocks in the market. Investors, be they individuals, pension, funds, or other corporations, will buy these stocks and allow the corporation to grow. There are ups and downs in the market. Generally, investing in the stock market is profitable for the careful investor. Some corporations are closed. In other words, they sell stock to only a certain small number of people, such as the organizers of the company or family members. For example, the Ford Motor Company was a closed corporation until 1956. More recently, many Internet companies were closed for years before they had an initial public offering (IPO). An IPO is when a corporation offers stock to the general public. Google's stock went up very quickly after its IPO. There are still many closed corporations, but many are publicly owned.

THE STOCK MARKET

There are many corporations in the nation (some say in the 2 million range). There are a great many more corporations worldwide. Many investors have invested in international, as well as American, corporations. A person or an organization who purchases stock in a company becomes a **stockholder** and is a part owner of the corporation. There are two main kinds of stock, **common** and **preferred**. Common stock gives voting rights to its holder. However, if there is a dividend to be paid, owners of preferred stock get paid before common stockholders. Preferred stockholders get paid first if the company goes bankrupt, but they generally do not have voting rights in the corporation. A **stock certificate** is a document that proves the holder is indeed the owner of stock in a corporation.

When the directors of a corporation have profits to pay out, these profits are called **dividends**. They are paid out to stockholders. When a stock is originally issued, its original selling price is called the **offer**. On the other hand, a **bid** is the highest price anyone is willing to pay for the stock. A **market order** is an order for a stockbroker to buy (or sell) a certain number of shares of stock at the best available price. A **round lot** of shares is a number of shares in a unit or a multiple thereof. In most markets, the unit is 100 shares; in some markets, it is 10. An **odd lot** is a number of shares of stock less than a round lot.

THE STOCK EXCHANGE

The stock exchange is a marketplace where approved stocks can be bought and sold. Stock exchanges are in New York, Chicago, Los Angeles, and other major cities. More and more, stock exchanges in other countries are important to stockholders here in America. Newspapers publish daily information about stock exchanges both here and around the world. However, the computer offers investors a much more immediate venue for keeping up with stocks.

UNDERSTANDING STOCK MARKET REPORTS

The newspaper stock market reports usually contain several vital pieces of information:

From: *THE NEW YORK TIMES*, Wednesday May 24, 2000

NEW YORK STOCK EXCHANGE

Continued From Preceding Page

52-Week High	52-Week Low	Stock	Div	Yld %	P/E	Sales 100s	High	Low	Last	Chg
14	5 1/2	JilinCh	.12e	1.8	3	51	6 3/4	6 1/2	6 1/2	- 1/2
17 1/4	8	JoAnnSt A	...	...	6	311	9	9	9	+ 1/16
13 1/8	6	JoAnnSt B	...	...	6	94	8	7 15/16	7 15/16	...
22 5/16	**13 7/16**	**JHFnSrv n**	...	...	...	**10073**	**21 3/4**	**20 1/2**	**21 3/4**	**+1**
43 11/16	33 1/2	JNuveen	1.04	2.5	14	429	41 7/8	41 1/8	41 3/8	+ 3/16
15	7 7/16	JohnsMnv	.24	2.0	7	549	12	11 3/4	11 13/16	- 5/16
106 7/8	66 1/8	JohnJn	1.28f	1.5	33	28055	88 3/16	86 1/8	87	+ 1/8
73 1/8	46 7/8	JohnsnCtrl	1.12	1.8	13	3119	61 1/8	60	60 3/4	+ 3/8
35 7/16	20 1/8	JonesApp	...	...	17	3133	28	27 3/4	27 15/16	+ 7/16
32	9 3/16	JonesLL	...	...	dd	388	15 1/16	15	15	- 1/16
23	11 7/16	JrnlReg	...	...	13	197	14 3/16	13 13/16	13 13/16	- 7/16
		K								
10 9/16	6 3/16	K2 Inc	...	...	12	480	7 19/32	7 5/16	7 9/16	+ 3/16
2 3/8	3/16	vjKCS	...	...	6	1096	1 1/4	15/16	1 1/16	+ 1/8
29 1/16	16 3/4	KLM n	...	...	...	485	21 1/16	20 1/2	20 1/2	...
17 1/2	7 5/16	K mart	...	...	11	18466	8	7 3/4	7 15/16	+ 1/8
47 1/2	23 3/8	KN En01	3.55	7.9	...	508	44 15/16	44	44 15/16	+ 9/16
143 7/16	39 11/16	KPN	1.04e	1.4	...	409	77 15/16	76 1/16	76 1/8	-2 5/8
32 11/16	14 3/4	KV Ph B	...	...	18	45	22 3/8	22	22	- 1/4
32 5/8	14 3/8	KV Ph A	...	...	18	133	22	21 1/2	21 3/4	+ 1/8
9 11/16	2 15/16	KaisAl	...	...	dd	748	4 9/16	4 3/8	4 3/8	+ 1/8
32 5/8	22	KanPipLP	2.80	11.3	11	117	24 15/16	24 3/4	24 7/8	- 1/16
6 15/16	4	Kaneb	...	...	3	171	5 3/16	5 1/16	5 1/8	- 1/16
29	20 13/16	KCtyPL	1.66	6.9	22	1354	24 3/8	23 5/8	24 3/16	+ 1/8
93 7/8	37 1/2	KC Sou	.04j	...	21	3272	70	68 1/8	68 7/16	-1 13/16
17 1/4	**7 5/8**	**KatyInd**	**.30**	**2.7**	**10**	**110**	**11 1/2**	**11**	**11**	**- 3/4**
25 9/16	16 3/4	KaufBH	.30	1.7	6	2186	17 7/8	17 7/16	17 7/16	- 1/16
34 3/4	21 9/16	Kaydon	.44	1.9	13	352	23 5/8	22 3/4	22 7/8	- 3/16
36 1/8	21 3/4	Keebler	.11p	...	29	1722	35 9/16	34 1/2	34 7/8	- 1/16
74	**8 3/16**	**Keithly**	**.22f**	**0.4**	**27**	**1791**	**62 1/8**	**56 3/16**	**56 1/4**	**-3 9/16**
40 15/16	20 3/4	Kellogg	.98	3.2	32	13479	30 3/4	29	30 7/16	+1 1/8
27 13/16	13 3/4	Kellwood	.64	3.9	49	555	16 11/16	16 1/4	16 1/4	- 3/16
88 7/16	**16 3/4**	**Kemet**	...	...	**38**	**9752**	**76 3/4**	**63 1/2**	**64 3/8**	**-12 1/2**
9 5/16	7 7/16	KmpHi	.97	12.3	q	445	7 7/8	7 13/16	7 7/8	+ 1/16
7 3/16	6 1/8	KmpIGv	.54	8.5	q	1044	6 3/8	6 1/4	6 3/8	+ 1/8
9 3/8	7 11/16	KmpMI	.93a	11.3	q	466	8 1/4	8 1/16	8 1/4	+ 1/8
12 15/16	9 7/16	KmpMu	.82	7.7	q	802	10 13/16	10 5/8	10 11/16	- 1/16
17 3/8	12 1/8	KmpSInc	1.80	13.2	q	46	13 7/8	13 5/8	13 5/8	- 5/16
12 1/16	9 3/8	KmpStr	.75	7.3	q	[illegible]	10 5/16	10 1/8	10 5/16	+ 3/16
33 7/8	22 5/8	Kennmtl	.68	2.5	17	1746	27 3/8	26 11/16	26 15/16	...
46 1/8	17 5/8	KCole s	...	...	29	479	37 3/4	36 1/2	37 1/2	+ 1/2
42	**12**	**KentEl**	...	...	**43**	**1577**	**28 1/8**	**26 7/8**	**26 7/8**	**-1 3/8**
67 15/16	39 7/8	KerrMc	1.80	3.1	12	3511	59 13/16	58 1/2	58 1/2	- 3/8
53 1/4	30 13/16	KerrM04 n	1.83	3.5	...	2215	52 1/2	51 1/2	52 1/8	- 1/16
12 1/4	3	KeyEng	...	...	dd	2574	11 7/16	11	11 3/16	+ 1/16
15 3/8	6 3/4	KeyPrd	...	...	16	735	14 5/16	13 1/4	13 5/8	- 3/8
36	15 9/16	Keycorp	1.12	5.6	8	18418	20 9/16	19 1/4	19 7/8	+ 11/16
31 1/16	20 3/16	Keyspan	1.78	6.2	15	2202	29 3/16	28 13/16	28 7/8	...
7 11/16	3 3/4	KeyCon	...	...	dd	35	4	3 13/16	4	...
28 1/2	18	KilroyR	1.80f	7.8	16	808	23 3/4	23 1/16	23 1/8	- 3/8
69 9/16	42	KimbClk	1.08	1.8	18	10356	60 1/16	57 3/4	59 3/4	- 1/4
42 1/4	30 7/8	Kimco	2.64	6.4	16	466	41 1/2	40 3/4	41 1/8	+ 5/16
45 5/8	35 1/4	KindME	3.10f	8.2	14	1330	38 1/4	37 1/4	37 5/8	- 5/16
34 1/2	12 3/16	KindMorg	.80	2.6	15	2578	31 1/4	30 3/8	30 15/16	+ 1/8
68 3/4	13 7/8	KingPh s	...	...	78	1525	51 3/4	51	51 1/2	...
2 3/4	1 1/8	Kinross g			...	7512	1 3/16	1 1/16	1 1/16	- 1/8
24 5/8	16 1/2	Kirby	...	...	21	147	23	22 3/8	22 1/2	- 1/2
66	44 7/16	KnightR	.92	1.8	11	3819	51 15/16	50 5/16	51	+ 3/8
19 3/16	14 1/4	KogerEq	1.40	7.9	14	100	17 3/4	17 3/8	17 11/16	+ 5/16
54 3/4	30 3/4	Kohls s	...	...	65	18277	53 1/2	52 3/16	52 5/8	+1 3/8
22 7/8	8 7/8	[illegible]	.08	0.4	20	196	22 3/4	22 11/16	22 11/16	
9 3/16	4 3/4	Konover	.50	10.3	dd	51	[illegible]	4 7/8	4 7/8	[illegible]
24 3/4	16	Koor	.89e	3.9	dd	61	18 1/4	17 3/4	17 3/4	- 5/8
21 3/4	11 3/4	KoreaElc	.27e	1.7	...	2190	15 9/16	14 13/16	15 1/2	+ 11/16
5 15/16	3 1/16	KoreaEqt	...	...	q	303	3 3/16	3 1/8	3 3/16	- 1/16
17 13/16	11 1/8	Korea	...	...	q	1811	11 7/16	11 3/16	11 3/16	- 5/16
79 7/8	30 1/16	KoreaTel	.20e	0.6	...	6094	34	33 1/4	33 3/4	+ 1/4
9 1/8	5 1/2	KoreaInv	...	...	q	1393	6	5 5/8	5 15/16	- 1/16
44 1/8	12 1/2	KornFer	...	...	...	212	22 3/4	22 5/16	22 1/2	...
14 1/8	7 1/2	Kranzc	1.30	14.3	22	197	9 1/4	8 15/16	9 1/16	- 1/16
31 3/8	14 1/16	Kroger s	...	...	27	33536	20 1/4	19 9/16	19 9/16	- 7/16
86 1/2	54	Kubota	1.13e	1.5	31	5	74	73	74	+2
280 9/16	51 1/16	Kyocera	.57e	0.4	...	277	144 1/4	140 3/4	141 3/4	+1 1/4
		L								
58 3/4	**34 1/4**	**L-3 Com**	...	...	**30**	**2006**	**55 3/4**	**51 15/16**	**55 1/4**	**+3 1/16**
23 13/16	15 1/4	LG&E	1.27	5.4	15	2522	23 3/4	23 5/8	23 11/16	+ 1/8
3 3/4	3/4	LLE Ry	.24e	6.9	7	353	3 5/8	3 7/16	3 1/2	- 1/16
24	16 1/2	LNR Pr	.05	0.3	7	51	19 3/16	18 7/8	18 7/8	- 1/16
[illegible]	**[illegible]**	**LSI Log s**	...	...	**55**	**31544**	**[illegible]**	**42**	**42 5/8**	**-2 3/8**
13 7/8	4	LTC Prp	1.16m	20.0	11	615	6	5 13/16	5 13/16	...
7 7/16	2 5/8	LTV	.12	3.9	dd	2759	3 1/8	3	3 1/16	- 1/16
24 9/16	13 1/2	LaZBoy	.32	2.1	10	549	15 1/4	15 1/16	15 1/16	+ 1/16
28 3/8	7 5/8	LabrRdy s	...	...	24	617	9 11/16	9 7/16	9 11/16	...
25 1/2	15 7/8	LabChile	1.03e	5.3	...	22	19 3/4	19 9/16	19 9/16	- 3/16
74 5/8	22 1/2	LabCp s	...	...	39	1347	67 7/16	65 3/8	66 1/2	+ 3/8
15 3/4	9 1/16	LaBrnch n	...	...	7	563	11 3/4	11 1/2	11 3/4	...
23 3/4	17 1/2	LaclGas	1.34	6.8	15	132	19 5/8	19 3/8	19 9/16	+ 3/16
36 7/8	18 13/16	Lafarge	.60	2.4	7	1310	26 5/8	25 1/16	25 3/16	- 11/16
8	7/16	Laidlaw	.07j	...	dd	11914	9/16	15/32	1/2	- 1/16
45 15/16	32	LakehdP	3.50	9.3	16	419	38 1/8	37 9/16	37 3/4	- 1/4
8 3/4	4 5/16	LamSes	...	...	5	19	8	8	8	+ 1/8
30 1/2	15 9/16	LandAmer	.20	1.1	10	143	18 1/2	18 1/16	18 1/16	- 7/16
10 3/16	6 1/16	Landrys	.03p	...	dd	769	8 1/2	8 3/8	8 7/16	...
83 1/2	27 1/4	LandsE	...	...	14	1319	33 11/16	32 9/16	32 13/16	- 3/8
6 1/2	1 5/16	LanierW n	...	...	...	2154	1 9/16	1 7/16	1 7/16	- 1/16
16 1/8	10 13/16	LaSalleH	1.52	11.0	16	243	13 15/16	13 5/8	13 13/16	+ 1/8
18 5/8	10 7/8	LaSalleRe	...	...	dd	59	13	12 7/8	13	...
4 1/16	3 1/8	LaserMtg	...	...	dd	17	3 13/16	3 3/4	3 13/16	...
8 3/16	4 1/8	Lasmo	.63e	11.2	80	88	5 11/16	5 3/8	5 5/8	- 1/16
14 5/16	8 1/4	LatAEqt	...	...	q	118	10 5/8	10 1/2	10 9/16	- 1/16
13 7/16	7 1/8	LatADisc	.03e	0.3	q	436	8 15/16	8 1/2	8 7/8	+ 1/8
14 5/16	9 9/16	LatAInv	.50e	...	q	34	12 11/16	12 5/8	12 5/8	- 1/8
53	19 1/4	LearCorp	...	...	6	2078	26 15/16	25 1/4	25 1/4	- 3/4
32 1/4	19 11/16	LeeEnt	.64	3.0	12	401	21 1/8	20 11/16	21 1/16	+ 3/8
51 1/4	30 5/8	LeggMas	.26	0.6	18	1004	43 1/16	41 1/16	41 15/16	+ 9/16
28 5/16	15 1/4	LeggPlat	.40f	2.0	13	2841	19 7/8	19 1/2	19 11/16	+ 1/16
107 1/2	47 9/16	LehmBr	.44f	0.6	10	7902	82 3/4	78 7/16	78 7/8	-2 13/16
25 7/8	21 5/8	LehBr35	2.08	9.2	...	102	22 5/8	22 1/2	22 5/8	...
24 15/16	13 1/16	Lennar	.05	0.3	7	957	17 7/8	17 7/16	17 3/4	- 1/4
19 7/8	7 7/8	Lennox n	.38	3.3	7	1199	11 1/2	10 7/8	11 7/16	+ 1/2
25 3/8	20 1/8	LeucNtl s	1.58e	7.0	15	283	23 1/4	22 5/8	22 11/16	+ 1/16
12 7/8	8 13/16	LexCrpP	1.20	11.2	10	97	10 3/4	10 9/16	10 11/16	+ 1/16
135 7/8	**59 5/8**	**Lexmark**	...	...	**30**	**28861**	**80 1/2**	**71 15/16**	**72 3/4**	**-8 1/4**
33 3/4	24 5/8	Libbey	.30	1.0	10	114	29 5/16	29	29 1/16	- 1/4
3 13/16	3	Liberte	.06e	2.0	50	7	3 1/8	3	3	...
14 1/8	9 3/4	LbtyASE	1.38e	12.4	q	1329	11 1/4	11 1/16	11 1/8	...
11 9/16	9 5/16	LbtyASG	1.24e	12.3	q	604	10 1/4	10 1/16	10 1/16	- 1/8
54 9/16	30 3/4	LibtyCp	.88	2.6	12	153	34 1/8	33 1/2	33 5/8	- 1/8
29 11/16	17 7/8	LibFin	.40	1.8	10	[illegible]	22 1/4	21 11/16	21 7/8	- 5/16
61 7/16	30 7/8	LibtyMA s	...	...	...	37900	40 1/4	38 3/4	39 11/16	- 1/8
73 1/8	32	LibtyMB s	...	...	cc	8	45 1/4	45	45 1/4	- 3/4
28 7/8	20 7/8	LibtProp	2.08	8.2	13	835	25 1/2	25 5/16	25 5/16	...
80 1/2	54	LillyEli	1.04	1.4	28	17110	76 3/16	73 7/8	75 9/16	+ 3/8
19 3/4	10 1/2	LillyInd	.32	2.3	10	552	14 7/16	13 7/8	13 7/8	- 1/4
51 3/4	28 7/8	Limitd	.60b	1.2	23	6309	49 11/16	48 5/16	48 3/4	- 1/2
21	12 3/4	LncNtC	.80	4.9	q	71	16 1/4	16 1/8	16 1/4	+ 1/4
57 1/2	22 5/8	LincNat s	1.16	3.2	17	5704	36 7/8	34 7/8	36 3/8	+1 1/8
12 7/8	9 9/16	LincNIF	.96a	9.2	q	43	10 7/16	10 7/16	10 7/16	- 1/16
21 1/4	13 5/8	Lindsay	.14	0.7	18	37	18 3/4	18 7/16	18 11/16	+ 5/16
49 1/2	17 15/16	Linens	...	...	20	2006	30 9/16	30	30 5/16	- 1/8
23 3/4	11 5/8	LithiaMot	...	...	7	253	12 15/16	12 3/8	12 1/2	- 1/8
74 5/8	26 13/16	Litton	...	...	16	3568	42 3/8	40 5/8	40 7/8	- 1/8
48 5/16	30 15/16	LizClab	.45	1.1	13	2387	42 5/16	41 13/16	42 1/16	+ 3/8
42 1/4	**16 3/8**	**LockhdM**	**.44m**	**1.8**	**18**	**11093**	**25 5/16**	**23 5/8**	**25 1/8**	**+1 9/16**
7 1/8	2 3/16	▼Lodgian	...	...	dd	1221	2 1/16	2	2 3/16	...
11 1/8	2 1/4	LoewsCin	...	...	dd	86	2 13/16	2 5/8	2 5/8	...
82 1/2	38 1/4	Loews	1.00	1.6	13	3801	65 1/16	63 1/4	64	+ 3/4
32 3/4	[illegible]	LondnPc s	.23e	[illegible]	...	[illegible]	10 9/16	10	10 1/8	- 3/8
53 5/16	14 7/16	LoneStTch	...	...	cc	279	49 9/16	48 3/4	48 7/8	- 13/16
36 7/8	15 15/16	LongDrg	.56	2.8	12	417	20 11/16	20 1/4	20 1/4	- 3/16
[illegible]	[illegible]	Longvw	.32e	2.7	18	[illegible]	12 1/2	11 13/16	11 7/8	- 9/16
25 3/4	7 9/16	LoralSp	...	...	dd	18904	8 7/16	7 15/16	7 15/16	...
34	15 3/4	LDryNG	...	...	38	1613	33 1/4	32 5/8	33 1/16	+ 3/8
24 7/8	10 13/16	LaPac	.56	4.3	6	4873	13 5/16	12 13/16	12 7/8	- 5/8
67 1/4	40 3/4	Lowes	.14	0.3	25	10021	49 5/8	48	48 5/16	- 9/16
33 7/8	23	Lubrizol	1.04	4.2	12	1285	25	24 3/8	24 1/2	- 3/8
17 1/4	8 3/4	Lubys	.80	7.8	8	522	10 7/16	10 1/8	10 5/16	+ 1/16
84 3/16	49 13/16	Lucent	.08	0.1	49	98199	56 3/8	54 9/16	54 9/16	+ 1/16
27	13 7/8	Luxottc	.16e	0.7	34	207	24 1/2	24 1/8	24 7/16	+ 3/16
12 7/8	5 1/8	Lydall	...	...	12	419	10 5/16	10 3/16	10 1/4	...
22 1/2	8 7/16	Lyondell	.90	4.9	12	x4806	18 7/8	18 1/4	18 5/16	- 3/16
		M								
575	357	M&T Bank	5.00	1.2	13	z15180	424	410	422	+11 5/8
68 7/8	36 5/16	MBIA	.82f	1.4	13	4249	57 1/8	55 1/4	56 5/8	+ 3/4
24 15/16	19 3/8	MBIA38	1.74	8.5	...	13	20 13/16	20 9/16	20 9/16	...
33 1/4	19 1/2	MBNA	.32	1.2	22	21186	27 15/16	26 15/16	27 7/16	+ 11/16
26 15/16	**17**	**MCN Engy**	**1.02**	**4.8**	**dd**	**6174**	**22 15/16**	**21 7/16**	**21 7/16**	**-1**
14 1/4	4 7/8	MSC Sft	...	...	15	315	8 3/4	8 5/16	8 3/4	+ 1/8
22	13 3/8	MDC	.24	1.0	4	[illegible]	19 7/16	18 7/8	19	+ 1/16
32	25 1/2	MDS gn	.07e	...	...	8	31 1/4	30 7/8	30 7/8	- 1/8
24 3/4	17 5/8	MDU	.84	4.0	14	347	22	21 1/4	21 1/4	- 5/8
24 1/4	**7 3/4**	**MEMC**	...	...	**dd**	**2275**	**14 7/8**	**13 3/16**	**14**	**+ 7/8**
9 3/16	7 3/4	MCR	.79	10.0	q	1025	7 15/16	7 13/16	7 15/16	+ 1/16
6 5/16	5 5/8	MGF	.52	8.9	q	643	5 7/8	5 13/16	5 7/8	+ 1/16
6 1/2	5 15/16	MIN	.55	9.1	q	1289	6 1/8	6	6 1/16	- 1/16
6 3/4	5 1/2	MMT	.59	10.4	q	915	5 13/16	5 11/16	5 11/16	- 1/8
8 1/2	6 3/8	MFM	.53	7.5	q	365	7 1/16	7	7 1/16	+ 1/16
17 11/16	13	MFV	1.65	12.6	q	54	13 1/2	13 1/8	13 1/8	- 3/8
9 15/16	2 1/4	MGI Prp s	26.50c	...	2	425	2 3/8	2 1/4	2 1/4	- 1/16
62 3/4	31 15/16	MGIC	.10	0.2	11	3932	49 7/8	47 9/16	49	+ 5/8
35 1/4	18 7/16	MGM Gr s	...	...	31	4140	33 7/16	32 7/8	33	- 1/8
20 3/16	12 3/4	MISchott	.20	1.2	4	26	16 3/4	16 11/16	16 11/16	- 3/16
18 5/16	9 7/8	MIIX n	.20	1.8	7	37	11 3/8	11 3/8	11 3/8	- 1/16
5 1/4	3 1/2	ML Macad	.50	10.3	9	56	4 7/8	4 13/16	4 7/8	+ 1/16
22 3/16	7 1/2	MSC Ind	...	...	31	851	21 1/4	21 1/16	21 1/16	+ 1/16
24 15/16	19 3/16	MSDW38	1.78	8.8	...	306	20 3/8	20 1/8	20 5/16	+ 1/16
46 3/4	19 3/4	McDrmd	.08	0.4	16	724	20 3/4	20 3/8	20 9/16	- 3/16
27 1/4	17 5/16	Macerich	2.04	9.2	19	555	22 15/16	22 3/16	22 1/4	+ 1/4
10	2 3/4	MacGry	...	...	dd	1218	3	2 13/16	2 13/16	- 3/16
33 3/16	22 3/4	MackCali	2.32	8.5	13	1661	27 1/2	27 1/16	27 3/8	+ 3/8
15 15/16	7 5/8	Madeco	...	...	6	109	8 1/4	8 1/8	8 1/8	- 1/4
10 1/2	1 3/4	Magellan	...	...	dd	802	2 5/16	2 3/16	2 1/4	...
62 3/4	38 3/16	MagnaI g	1.20	2.3	...	1735	52 7/8	49 7/8	51 1/4	+1 7/16
11 5/8	6 1/8	Magntk	...	...	7	277	8 1/8	8 1/16	8 1/16	+ 1/16
17 13/16	6 3/4	MailWell	...	...	7	511	9 3/8	9	9 3/16	- 1/8
15 7/8	11 7/8	MalanR	1.70	12.4	15	68	13 3/4	13 5/8	13 11/16	+ 3/16
7 13/16	5 1/8	Malaysa	...	...	q	94	5 1/4	5 3/16	5 3/16	- 1/16
37 9/16	22 3/16	Malinckr	.66	2.2	11	1781	30 13/16	30 1/4	30 5/16	- 1/16
10 9/16	8	MgdHi	1.01	12.4	q	993	8 3/16	8 1/8	8 1/8	...
11 7/16	9	MgHiYld	1.26	13.5	q	227	9 3/8	9 1/4	9 5/16	...
12 5/8	9 9/16	▼MgHiYP	1.50	16.0	q	2056	9 11/16	9 3/8	9 3/8	- 5/16
10 13/16	9	MgdMun	.60	6.5	q	1085	9 3/16	9 1/8	9 3/16	...
10 7/8	8 15/16	MgdMun2	.60	6.6	q	397	9 1/8	9 1/16	9 1/8	...
24 1/2	12 7/8	Mandalay	...	...	16	6564	22 3/8	21 3/4	22 1/16	+ 1/8
43 3/4	24 9/16	Manitow	.30	0.9	13	263	33 13/16	33 1/16	33 1/4	- 1/8
28 1/16	6 7/16	ManorCare	...	...	dd	1299	7	6 13/16	6 7/8	- 1/16
39 15/16	21	Manpwl	.20	0.6	34	2032	36 3/4	35 5/16	35 7/8	+ 3/8
26 15/16	22 1/4	ManufHm	1.66f	6.6	23	779	24 3/8	24 1/16	24 5/16	+ 3/16
16 11/16	10 1/4	▲Manulif gn	...	...	...	4590	17 1/2	16 3/16	17	+ 11/16
14 5/8	8 3/16	Marcus	.21	2.0	14	28	10 7/8	10 1/2	10 1/2	...
31 5/16	**11 5/16**	**MarineDrl**	...	...	**cc**	**5745**	**27 1/2**	**25 5/8**	**27 1/2**	**+1 13/16**
12 1/8	8 7/16	MarineMx	...	...	7	29	9 1/4	9 3/16	9 1/4	...
6 15/16	4 3/4	Maritm	.40	7.0	7	29	5 3/4	5 5/8	5 3/4	+ 1/8
23 7/8	16 1/2	MarkIV	.25f	1.1	13	614	22 1/16	21 13/16	21 13/16	...
192	111 1/2	Markel	...	...	20	175	141	140 1/8	140 7/8	+ 3/16
40 13/16	26 1/8	MarIntA	.24f	0.7	22	2479	36 3/16	35 7/16	35 5/8	- 9/16
110 11/16	61 3/4	MarshMcL	2.00f	1.9	38	4979	106 1/4	104	104 15/16	-1 5/16
70	42 3/4	Marshlls	1.06f	2.2	15	2087	49 7/8	48 11/16	49	- 1/16
47 1/2	13 1/16	MStewrt n	...	...	54	232	17 15/16	17 7/16	17 7/16	- 5/16
64 1/2	35 1/4	MartMM	.52	1.0	19	1260	52	50 15/16	51 1/2	- 3/4
9 1/8	4 1/16	MarvelEnt	...	...	dd	1124	4 5/16	4 3/16	4 1/4	+ 1/16
33 11/16	17 1/16	Masco	.48	2.4	15	17517	20 5/16	19 9/16	20 3/16	+ 3/8
17 11/16	10 5/8	MascoTch	.32	2.6	7	1001	12 11/16	12 5/16	12 7/16	+ 1/8
23 3/4	17 5/8	MassCp	1.72a	8.0	q	53	21 7/16	20 15/16	21 3/8	+ 3/8
11 3/4	8 1/2	MasPrt	.96a	9.8	q	51	9 3/4	9 5/8	9 3/4	+ 1/16
90 1/4	**21 1/2**	**Mastec**	...	...	**35**	**2534**	**70**	**[illegible]**	**[illegible]**	**[illegible]**
49 15/16	25 13/16	Matav	.15e	0.5	...	x748	29 1/2	28 15/16	28 15/16	+1 1/16
16	9 1/2	MatSci	...	...	9	142	10 5/8	10 1/4	10 1/4	1/4
5 15/16	1 7/16	Matlack	...	...	dd	108	1 3/4	1 5/8	1 3/4	+ 1/4
303	173 1/2	Matsu	1.17e	0.5	76	37	235 5/16	231	231	+1
[illegible]	[illegible]	Mattel	[illegible]	[illegible]	dd	[illegible]	[illegible]	[illegible]	[illegible]	[illegible]
3 7/8	2 5/16	Mavesa	.08e	3.0	7	1353	2 3/4	2 5/8	2 5/8	...
38 15/16	20 3/4	Maximus	...	...	15	308	22 1/4	21 7/8	22 1/16	- 1/16
44 3/4	23 3/4	MayDS	.93	3.1	11	10934	29 15/16	29 5/16	29 5/8	+ 1/16
74 13/16	25 15/16	Maytag	.72	2.0	10	3336	36 9/16	34 11/16	35 13/16	+1 1/4
45 1/8	28 3/4	McClatchy	.40	1.3	16	198	29 7/8	29 1/2	29 13/16	+ 1/8
36 9/16	23 3/4	McCorm	.76	2.3	22	1112	33 5/8	33 5/16	33 7/16	...
30 3/16	7 1/4	McDerl	.20	2.1	67	1921	9 7/16	9	9 3/8	+ 1/8
49 9/16	29 13/16	McDnlds	.20	0.5	27	24521	39 1/4	37 3/4	39 1/8	+ 1/4
26 1/16	21 5/8	McDn36	1.88	8.2	...	76	23 1/16	22 7/8	23 1/16	...
26	21 3/4	McDn37	1.87	8.2	...	55	22 15/16	22 3/4	22 15/16	+ 1/8
60 1/8	41 7/8	McGrH	.94	1.9	22	3965	50 1/4	49 1/16	49 13/16	+ 5/16
39	16 1/8	McKHBOC	.24	1.5	6	8214	17	16 3/8	16 1/2	+ 1/8
25	11 5/8	McMoRn	...	...	dd	1202	13 3/8	12 3/4	13 3/8	+ 1/2
19 3/4	10 7/8	McWhrtr	...	...	21	437	19 5/8	19 7/16	19 1/2	+ 1/16
46 5/16	**28 1/8**	**Mead**	**.68**	**2.1**	**21**	**10919**	**33 3/4**	**31 3/4**	**32 1/16**	**-1 11/16**
14 1/16	4 3/4	Mdbklns	.12	2.4	dd	73	5 1/8	5 1/16	5 1/16	- 3/16
10 1/2	5 1/2	MedPtTAP	1.44	17.7	...	159	8 1/4	7 15/16	8 1/8	+ 1/16
9 1/4	3 1/16	MediaArt	...	...	4	125	4 5/8	4 5/8	4 5/8	...
84	64	MediaOne	...	...	9	21657	67 1/16	65 5/16	65 83/128	- 125/128
135	78 1/2	MediaO 01	1.56	1.8	...	22	84 1/2	81 1/2	84 1/2	+2 1/2
69 1/2	36 3/4	Media02 n	3.04	7.8	...	1005	39 7/8	38 1/16	39	+ 5/8
29	10	MedAsr	...	...	6	459	11 1/2	10 7/8	11 1/4	+ 3/8
51 15/16	20 1/4	Medicis	...	...	29	1562	46 1/2	44 1/8	45 7/8	+1 1/2
14 1/4	1 13/16	Meditrust	...	...	4	1777	2	1 15/16	1 15/16	- 1/16
57 7/8	**29 15/16**	**Medtrnic s**	**.16**	**0.3**	**61**	**67987**	**49 1/2**	**46 1/16**	**47 7/8**	**-2 7/16**
40 3/16	26 13/16	MellonFnc	.88f	2.4	19	18458	37	35 7/16	36 1/8	+ 15/16
8 3/8	6 15/16	MentInc	.72	10.0	q	490	7 1/4	7 1/8	7 3/16	+ 1/16
81 1/8	52	Merck	1.16f	1.6	29	47473	73 13/16	72 5/8	73 13/16	+1 1/2
37 1/8	20 15/16	MercGn	.96	3.6	12	432	26 5/8	25 3/4	26 9/16	+ 11/16
42	**22 3/8**	**Meredith**	**.32**	**1.1**	**17**	**954**	**29 3/16**	**27 9/16**	**29 1/8**	**+1 7/16**
9 9/16	3 13/16	MeridGld	.05	0.8	dd	393	6 1/4	5 15/16	6 1/8	+ 1/8
6 1/8	2 9/16	MeridRes	...	...	9	767	4 15/16	4 3/4	4 3/4	- 1/16
24 1/16	14 7/16	MeriStHsp	2.02	10.7	8	1185	19 1/8	18 7/8	18 15/16	- 1/8
4 1/2	2 1/8	MeriStHR	...	...	11	350	2 7/8	2 3/4	2 13/16	...
13 1/2	8 3/4	MeritagCp	...	...	3	39	11 11/16	11 1/2	11 11/16	+ 1/8
26 1/2	12 3/4	Meritor	.42	3.2	3	1760	13 1/4	12 15/16	13	+ 1/16
115 3/8	62	MerrLyn	1.20f	1.2	14	25480	103 3/8	99 3/8	99 3/8	-2 7/16
54 1/2	34	MLCBR01	4.26	12.4	...	87	35	34	34 3/8	+ 3/16
13 3/16	11 7/16	ML DJ03	...	...	...	6	12 1/8	12 1/8	12 1/8	...
25	14 1/2	MLIGL01	2.39	13.2	...	129	18 1/4	18	18 1/8	+ 1/8
10 5/8	9	ML Nik02	...	...	...	20	9 1/8	9 1/8	9 1/8	- 1/16
24 15/16	19 7/8	MLSP01	...	...	...	88	22 9/16	22 1/2	22 1/2	- 3/16
18 5/8	15	MLSP02	...	...	...	34	16 3/4	16 11/16	16 11/16	- 1/16
12 1/8	10 1/4	MLSP905	...	...	...	28	11 3/8	11 1/8	11 1/8	- 3/16
11	9	MLSP07	...	...	...	8	10 1/2	10 1/2	10 1/2	- 1/8
18 1/16	15 3/4	ML Tech01	...	...	...	11	17 3/4	17 3/4	17 3/4	+ 1/16
3 3/8	2 3/16	Mesab	.38e	13.2	...	70	3 1/16	2 7/8	2 7/8	- 1/16
23	14 3/16	Mestek	1.31t	...	8	72	16 3/4	16 1/2	16 1/2	...
13 5/8	8 11/16	MetPro	.32	3.6	8	1144	8 15/16	8 11/16	8 13/16	+ 1/16
13 1/2	6 3/16	Metals	.03p	...	6	287	6 1/2	6 5/16	6 5/16	- 3/16
19 11/16	14 5/16	MetLife n	...	...	...	20939	19	18 1/4	18 7/8	+ 5/8
64 1/4	50	MetLife un	...	...	...	2116	62 1/4	60 9/16	62	+1 1/16

DIVIDENDS DECLARED

Company	Period	Rate	Stk of record	Payable
IRREGULAR				
Coflexip Spons ADR	x	.5307	6-6	6-27
x- approx amount.				
Delta-Galil IndADS	x	.0931	5-25	6-12
Gabelli EquityTr	...	.27	6-16	6-26
Liberty All Star	...	.36	6-2	6-19
Liberty All-StarGwth	...	.33	6-2	6-19
LUMH Moet ADR	x	.4392	6-5	6-26
x- approx amount.				
Mesa Offshore	...	.0018	5-31	7-31
Mesa Royalty Tr	...	.2669	5-31	7-31
Supermercado Unimarc	...	.1581	5-30	6-9
Swiss Helvetia Fd	...	.045	5-31	6-7
Torch EnergyRoy	...	.342	5-31	6-12
CORRECTION				
Intel Corp	x	.02	8-7	9-1
x- increased amount payable on post split shares.				
Penn-American Grp	x	.0525	6-3	6-19
x- correcting pay date.				
Std Register	x	.23	5-26	6-9
x- correcting record date.				
INCREASED				
Gibraltar Steel	Q	.03		
x- dates unannounced.				
Home-Stake Oil	Q	.05	5-31	6-15
PAB Bkshrs	Q	.1075	6-30	7-14
Roslyn Bncp	Q	.15	6-2	6-13
FINAL				
Dominguez Svcs	...	.25	5-22	5-26
REDUCED				
JDN Realty	x	.30		
x- dates unannounced.				
Municipal Advantage	x	.06		
x- dates unannounced.				
REGULAR				
Advanta Cp A,	Q	.063	5-30	6-16
Advanta Cp B,	Q	.0756	5-30	6-16
Amer Bal Fd	M	.14	5-19	5-22
Becton Dickinson	Q	.0925	6-9	6-30
Blockbuster Inc	Q	.02	6-5	6-19
CenturyTel Inc	Q	.0475	6-2	6-16
Columbus McKinnon	Q	.07	6-22	7-6
Community Bk Syst	Q	.25	6-15	7-10
Corporate Office	Q	.19	6-30	7-17
Craftmade Intl	Q	.02	6-30	7-14
Doral Fncl A,	M	.2917	5-26	5-31
Edison Intl	Q	.28	7-5	7-31
Fst BanCorp PR,	Q	.11	6-15	6-30
Fst Fedl BcshrsArk	Q	.10	6-9	6-20
1st Midwest Fncl	Q	.13	6-15	7-3
Florida Progress	Q	.555	6-5	6-20
Gabelli Conv Secs	Q	.20	6-16	6-26
Gables Resident	Q	.53	6-16	6-30
Gulf CdaResadjpf1 g	Q	.024	5-31	6-12
HCB Bnshrs Inc AR,	Q	.06	6-15	6-29
HSBC Bk ADR A,	Q	.5546	5-26	6-15
HSBC Bk ADR B,	Q	.6406	5-26	6-15
HSBC Bk ADR C,	Q	.5703	5-26	6-15
HSBC Bk ADR D,	Q	.5968	5-26	6-15
Hartford Life Inc	Q	.10	6-1	7-3
Home City Fncl	Q	.105	6-5	6-15
Home Fedl Bncp	Q	.1375	6-20	7-3
Magna Intl A,	Q	.30	5-31	6-15
Martin Marietta	Q	.13	6-1	7-3
Merchants NY Bncp	Q	.125	6-15	6-29
NUI Corp	Q	.245	6-1	6-15
Natl City Bkshrs	Q	.21	6-21	7-7
Precision Castpart	Q	.06	6-2	7-3
Presidentl Rlty A,	Q	.16	6-9	6-30
Prosperity Bcshrs	Q	.09	6-16	7-3
Salisbury Bncp	Q	.13	6-30	7-28
Scientific-Atlanta	Q	.01	6-1	6-16
Sentinel Bd A,	M	.035	5-24	5-31
Sentinel Bd B,	M	.031	5-24	5-31
Sentinel Gvt Secs	M	.052	5-24	5-31
Sentinel PA TxFr	M	.052	5-24	5-31
Sentinel Sht/Mat	M	.053	5-24	5-31
Suffolk Bncp	Q	.23	6-15	7-3
UB Bancorp	Q	.215	6-1	6-15
US Indus	Q	.05	6-30	7-21
Unocal Corp	Q	.20	7-10	8-10
Westvaco Corp	Q	.22	6-2	7-3
g- payable in Canadian funds.				

52-Week High	52-Week Low	Stock	Div	Yld %	P/E	Sales 100s	High	Low	Last	Chg
14	5 1/2	JilinCh	.12 e	1.8	3	51	6 3/4	6 1/2	6 1/2	− 1/2
17 1/4	8	JoAnnSt A	...	...	6	311	9	9	9	+ 1/16
13 1/8	6	JoAnnSt B	...	...	6	94	8	7 15/16	7 15/16	...
22 5/16	**13 7/16**	**JHFnSrv n**	...	...	...	**10073**	**21 3/4**	**20 1/2**	**21 3/4**	**+1**
43 11/16	33 1/2	JNuveen	1.04	2.5	14	429	41 7/8	41 1/8	41 3/8	+ 3/16
15	7 7/16	JohnsMnv	.24	2.0	7	549	12	11 3/4	11 13/16	− 5/16
106 7/8	66 1/8	JohnJn	1.28 f	1.5	33	28055	88 3/16	86 1/8	87	+ 1/8
73 7/8	46 7/8	JohnsnCtrl	1.12	1.8	13	3119	61 1/8	60	60 3/4	+ 3/8
35 7/16	20 1/8	JonesApp	...	...	17	3133	28	27 3/4	27 15/16	+ 7/16
32	9 3/16	JonesLL	...	...	dd	388	15 1/16	15	15	− 1/16
23	11 7/16	JrnlReg	...	...	13	197	14 3/16	13 13/16	13 13/16	− 7/16
		K								
10 9/16	6 3/16	K2 Inc	...	...	12	480	7 19/32	7 5/16	7 9/16	+ 3/16
2 3/8	3/16	vjKCS	...	...	6	1096	1 1/4	15/16	1 1/16	+ 1/8
29 1/16	16 3/4	KLM n	...	...	...	485	21 1/16	20 1/2	20 1/2	...
17 1/2	7 5/16	K mart	...	...	11	18466	8	7 3/4	7 15/16	+ 1/8
47 1/2	23 3/8	KN En01	3.55	7.9	...	508	44 15/16	44	44 15/16	+ 9/16
143 7/16	39 11/16	KPN	1.04 e	1.4	...	409	77 15/16	76 1/16	76 1/8	−2 5/8
32 13/16	14 3/4	KV Ph B	...	...	18	45	22 3/8	22	22	− 1/4
32 5/8	14 3/8	KV Ph A	...	...	18	133	22	21 1/2	21 3/4	+ 1/8
9 11/16	2 15/16	KaisAl	...	...	dd	748	4 9/16	4 3/8	4 3/8	+ 1/8
32 5/8	22	KanPipLP	2.80	11.3	11	117	24 15/16	24 3/4	24 7/8	− 1/16
6 15/16	4	Kaneb	...	...	3	171	5 3/16	5 1/16	5 1/8	− 1/16
29	20 13/16	KCtyPL	1.66	6.9	22	1354	24 3/8	23 5/8	24 3/16	+ 1/8
93 7/8	37 1/2	KC Sou	.04 j	...	21	3272	70	68 1/8	68 7/16	−1 13/16
17 1/4	**7 5/8**	**KatyInd**	**.30**	**2.7**	**10**	**110**	**11 1/2**	**11**	**11**	**− 3/4**
25 9/16	16 3/4	KaufBH	.30	1.7	6	2186	17 7/8	17 7/16	17 7/16	− 1/16
34 3/4	21 9/16	Kaydon	.44	1.9	13	352	23 5/8	22 3/4	22 7/8	− 3/16
36 1/8	21 3/4	Keebler	.11 p	...	29	1722	35 3/16	34 1/2	34 7/8	− 1/16
74	**8 3/16**	**Keithly**	**.22 f**	**0.4**	**27**	**1791**	**62 1/8**	**56 3/16**	**56 1/4**	**−3 9/16**
40 15/16	20 3/4	Kellogg	.98	3.2	32	13479	30 3/4	29	30 7/16	+1 1/8
27 13/16	13 3/4	Kellwood	.64	3.9	49	555	16 11/16	16 1/4	16 1/4	− 3/16
88 7/16	**16 3/4**	**Kemet**	...	...	**38**	**9752**	**76 3/4**	**63 1/2**	**64 3/8**	**−12 1/2**
9 5/16	7 7/16	KmpHi	.97	12.3	q	445	7 7/8	7 13/16	7 7/8	+ 1/16
7 3/16	6 1/8	KmplGv	.54	8.5	q	1044	6 3/8	6 1/4	6 3/8	+ 1/8
9 3/8	7 11/16	KmpMl	.93 a	11.3	q	466	8 1/4	8 1/16	8 1/4	+ 1/8
12 15/16	9 7/16	KmpMu	.82	7.7	q	802	10 13/16	10 5/8	10 11/16	− 1/16
17 3/8	12 1/8	KmpSlnc	1.80	13.2	q	46	13 7/8	13 5/8	13 5/8	− 5/16
12 1/16	9 3/8	KmpStr	.75	7.3	q	308	10 5/16	10 1/8	10 5/16	+ 3/16
33 7/8	22 5/8	Kennmtl	.68	2.5	17	1746	27 3/8	26 11/16	26 15/16	...
46 1/8	17 5/8	KCole s	...	...	29	479	37 3/4	36 1/2	37 1/2	+ 1/2
42	**12**	**KentEl**	...	...	**43**	**1577**	**28 1/8**	**26 7/8**	**26 7/8**	**−1 3/8**
67 15/16	39 7/8	KerrMc	1.80	3.1	12	3511	59 13/16	58 1/2	58 1/2	− 3/8
53 1/4	30 13/16	KerrM04 n	1.83	3.5	...	2215	52 1/2	51 1/2	52 1/8	− 1/16
12 1/4	3	KeyEng	...	...	dd	2574	11 7/16	11	11 3/16	+ 1/16
15 3/8	6 3/4	KeyPrd	...	...	16	735	14 5/16	13 1/4	13 5/8	− 3/8
36	15 9/16	Keycorp	1.12	5.6	8	18418	20 9/16	19 1/4	19 7/8	+ 11/16
31 1/16	20 3/16	Keyspan	1.78	6.2	15	2202	29 3/16	28 13/16	28 7/8	...
7 11/16	3 3/4	KeyCon	...	...	dd	35	4	3 13/16	4	...
26 1/2	18	KilroyR	1.80 f	7.8	16	808	23 3/4	23 1/16	23 1/8	− 3/8
69 9/16	42	KimbClk	1.08	1.8	18	10356	60 1/16	57 3/4	59 3/4	− 1/4
42 1/4	30 7/8	Kimco	2.64	6.4	16	466	41 1/2	40 3/4	41 1/8	+ 5/16
45 5/8	35 1/4	KindME	3.10 f	8.2	14	1330	38 1/4	37 1/4	37 5/8	− 5/16

- The difference between the preceding day's last price and the current day's last price.
- The sales prices of the particular day: the highest price, the lowest price, and the price of the day's last sale.
- The volume of sales in round lots of 100.
- The price/earnings ratio (the current stock price divided by the company's last 12-month earnings).
- The percentage return of the annual dividends divided by the current price of the stock.
- The current annual dividend per share, based on the latest quarterly or semiannual declaration.
- The abbreviation of the company name.
- The highest and the lowest prices of the stock in the last 52 weeks.

Charts on pages 155 and 156. Copyright © 2000 by *The New York Times* Company. Reprinted by permission.

EXAMPLE 1

What was the performance of MetLife Insurance Co. (n)?

SOLUTION

Look up MetLife in the third column.

MetLife's high price was 19 ($19).
MetLife's low price was 18¼ ($18.25).

MetLife's closing (last) price was 18 7/8 ($18.88).

MetLife was up 5/8 from yesterday.

Note that MetLife is selling on the high side of the yearly fluctuation of 19 11/16 and 14 5/16.

BROKERAGE FEES

Stock exchanges are meeting places where members of the exchange can buy and sell securities (stocks). Individuals or organizations put in orders to stock exchange members or have their stockbroker put in their order to an exchange member.

Stockbrokers put in orders on behalf of their clients and charge a **commission** (also called a **brokerage fee**) for their services. Some states also charge a **transfer tax**, which the seller of the stock must pay. Since it is just a small amount (and since it is charged in only some states), we will not consider it here.

Commission rates are different for each brokerage house. There are discount brokerage firms, and different commission rates are charged if the amount of the stock transaction is large or small. The following table lists general stockbroker commissions:

Minimum Commission Rates for Brokers

(Per 100-Share Orders and Odd-Lot Orders of Less than 100 Shares)

Money Involved in the Order	*Minimum Commission*
$100 but under $800	2.0% plus $ 6.40
$800 but under $2,500	1.3% plus $12.00
$2,500 and above	0.9% plus $22.00
Odd Lot = $2 less	

Ray Shaheen had his stockbroker buy 300 shares of Johnson Controls at 60½.

a) What was the broker's fee?

b) What was the total cost of the shares, including the broker's fee?

SOLUTION

When looking at the commission rate table above, the calculation is on 100 shares. When we calculate the commission, we will need to multiply by the multiples of the unit, in this case, 3.

a) Cost of 100 shares $\Rightarrow$ 100 $\times$ 60.5 = $6,050.00

Brokerage fee (on 100 shares): = 0.9% $\times$ 6,050 + 22

= 0.009 $\times$ 6,050 + 22

= $54.45 + 22

= $76.45

$$\text{Brokerage fee (on 300 shares)}: = \$76.45 \times 3$$
$$= \$229.35$$

b) Cost plus brokerage fee on 100 shares: $= \$6,050 + \$76.45$
$$= \$6,126.45$$

Cost plus brokerage fee on 300 shares: $= \$6,126.45 \times 3$
$$= \$18,379.35$$

$18,379.35 is the value of the stock with the brokerage fee included

Tom Erb sold 500 shares of Lennox Ceramics, Inc., through his broker. He sold the shares for 11 1/4. After the brokerage fees were deducted, how much was the sale worth?

SOLUTION

As with the previous example, we calculate the brokerage fee on 100 shares and then multiply by 5.

Selling price of 100 shares: $100 \times 11.25 = \$1,125.00$

$$\text{Brokerage fee (on 100 shares)}: = 1.3\% \times 1,125 + 12$$
$$= 0.013 \times 1,125 + 12$$
$$= 14.63 + 12$$
$$= \$26.63$$

$$\text{Broker's fee (on 500 shares)}: = \$26.63 \times 5$$
$$= \$133.15$$

Selling price of 500 shares: $= \$1,125 \times 5 = \$5,625.00$

$$\text{Net proceeds of sale}: = \$5,625.00 - 133.15$$
$$= \$5,491.85$$

Art Keegan bought (through his broker) 75 shares of MGM stock at 33¼ . What was the total cost of the sale, including the broker's fee?

SOLUTION

This is an example of an odd lot sale, so we will use the odd lot fee of $2 less than the usual fee.

Cost of 75 shares: $75 \times 33.25 = \$2,493.75$

$$\text{Brokerage fee (on 75 shares)}: = 1.3\% \times 2,493.75 + 10$$
$$= 0.013 \times 2,493.75 + 10$$
$$= 32.42 + 10$$
$$= \$42.42$$

Cost of stock plus brokerage fee: $\$2,493.75 + \$42.42 = \$2,536.17$

EXERCISES

Exercise A   Rewrite each of the following stock prices in equivalent decimal form:

1. $21 $\frac{1}{4}$ 2. $48 $\frac{3}{8}$ 3. $63 $\frac{5}{8}$ 4. $36 $\frac{7}{8}$ 5. $19 $\frac{3}{4}$

Find the net cost of the following lots of stocks at the indicated prices. (Change each price to a mixed decimal form.)

6. 40 at $67 $\frac{1}{4}$ 7. 75 at $63 $\frac{1}{2}$ 8. 400 at $36 $\frac{1}{8}$

9. 450 at $23 $\frac{1}{4}$ 10. 175 at $53 $\frac{7}{8}$

Find the preceding day's closing prices in each of the following stock prices. (Change all fractions to decimal equivalents.)

11. $18 $- \frac{1}{2}$ 12. $62 $\frac{3}{4} - \frac{7}{8}$ 13. $41 $- \frac{7}{8}$

14. $22 $\frac{3}{4} - 1\frac{7}{8}$ 15. $62 $\frac{1}{4} - \frac{7}{8}$

Refer to the stock market quotations on page 155. Find the net price of 350 shares of each of the following stocks that were bought at the prices indicated.
(NOTE. The net totals do not include brokers' fees and other expenses.)

	Company	When Bought	Price per Share		Net Total Price	
16.	Lucent	Low	____	___	_____	___
17.	Kroger s	High	____	___	_____	___
18.	K mart	Last	____	___	_____	___
19.	MBNA	Last	____	___	_____	___
20.	LaZBoy	High	____	___	_____	___

Exercise B   Find the brokerage fee on each of the following sales of stock:

	Number of Shares	Price per Share	Brokerage Fee	
21.	200	$13\frac{1}{4}$	_____	____
22.	50	$15\frac{1}{2}$	_____	____
23.	600	$9\frac{5}{8}$	_____	____
24.	75	$93\frac{1}{4}$	_____	____
25.	63	$12\frac{3}{4}$	_____	____

Find the total cost, including brokerage fees, for the following sales of stock:

	Number of Shares Bought	Price per Share	Selling Price		Brokerage Fee		Total Cost	
26.	300	$23	_____	___	_____	___	_____	___
27.	500	$8\frac{3}{4}$	_____	___	_____	___	_____	___
28.	60	$88\frac{7}{8}$	_____	___	_____	___	_____	___

Find the net proceeds for each of the following sales:

	Number of Shares Sold	Price per Share	Selling Price		Brokerage Fee		Net Proceeds	
29.	300	$31\frac{3}{4}$	_____	___	_____	___	_____	___
30.	400	$28\frac{1}{2}$	_____	___	_____	___	_____	___

Word Problems

Exercise C   Refer to the stock market quotations on page 155 in solving the following problems:

31. Frank Greene bought 250 shares of JNuveen at the low price in the quotations.

    (a) How much did he pay for the stock?
    (b) How much did he save by not buying the shares at the high for the day?

32. Janice Frank bought 275 shares of Libbey at the last price as listed in the quotations.

    (a) How much did she pay for the stock?
    (b) How much money did she save by not buying the stock at the preceding day's closing price?

33. Mrs. Washington owned 350 shares of LabCp s, which she had bought at the previous 52-week low. If she sold the shares at the preceding day's closing price, how much profit did she make?

34. Mrs. Sanchez bought 400 shares of MStewrt n at the 52-week low and sold the shares at the 52-week high.

    (a) How much did the shares cost?
    (b) How much profit did she make?

Unit 4: Investments–Bonds

Stocks are an important support to our economy, and bonds are equally important. Bonds, like stocks, are issued by companies when they need money to build factories, expand into new markets, or raise funds too large for a lending institution. Like stocks, bonds are offered to the public. The difference between stocks and bonds is that bonds carry with them no ownership in the company, like stocks do. Bonds are also issued by local, state, and federal governments. Bonds are essentially an IOU, a promise to pay bondholders back the face value of the bond (called the **par value**) and a specified interest rate (called the **bond rate**). This interest rate is also known as the **yield** in the bond trading table on page 162. Bonds are usually issued in units of $1,000. The **market value** of a bond is the price of the bond from the time it is issued until its maturity. Depending on market conditions, the price could be higher or lower than its par value.

There are three kinds of bonds. **Corporate bonds** are issued by corporations, hospitals, and other organizations. **Municipal bonds** are issued by local, county, and state governments. **Treasury** (or **government**) **bonds** are issued by the federal government.

BOND MARKET REPORTS

Bond market sales appear in daily newspapers in much the same way that stocks do. The following table is a small section of the New York Stock Exchange corporate bond market:

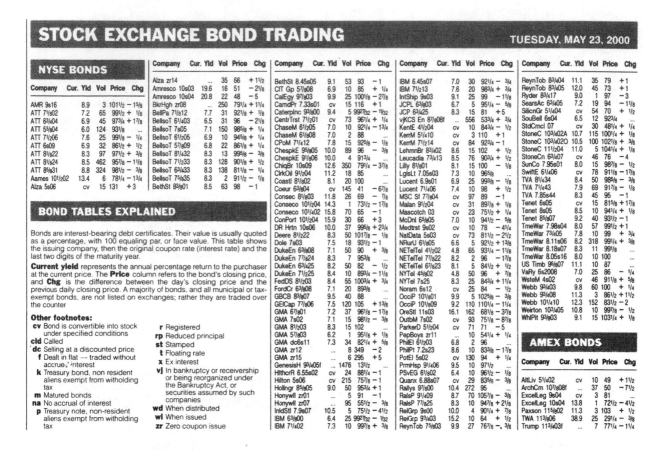

The bond market report is very much like the stock market report. Bonds, however, give a multiple of the *par value*, which is $1,000. Let's look at an example of this:

EXAMPLE 1

Find the market value of TVA (7 ¼ 43). It is listed at 91 7/8.

SOLUTION

The market value is 91 7/8% of $1,000. Since the quoted price is a percent of $1,000, we must first change to a decimal:

$$7/8 = 0.875$$
$$91\ 7/8 = 91.875$$

Since 91.875 is a percent, we must move the decimal two places to the left:

$$91.875\ \% = 0.91875$$

Finally, we multiply this number by 1,000:

$$0.91875 \times 1,000 = 918.75$$

The value of TVA (7¼ 43) bonds is $918.75.

Hint: Bond traders, knowing that the number first has to be divided by 100 and then multiplied by 1,000 (and hence the decimal place is moved first two places to the left and then three places to the right) simply move the decimal place one place to the right, which gives the same result.

 EXAMPLE 2   Find the net change on Ryder (8 ¾ 17). The listed chg is −3

SOLUTION

The column titled "chg" means *net change*. It is the difference between the stated price and the preceding day's closing price. In this case, the price has gone down 3 points.

$$-3 = -3.00$$

Using the hint from the previous example, we move the decimal place one place to the right: −30.0. The net loss from the previous day, is $30.

 EXAMPLE 3   Find the current yield (annual rate of return based on the current price) on PSvEG (6 1/8 02). The current price is 96½, and it pays an interest rate of 6.4%

SOLUTION

First, we calculate the cost of the bond: 96.5 ⇒ $965.00
Next, we calculate the interest paid on the par value:

Interest paid: = 6.4% × 1,000 = $64 (According to the hint in Example 1, we move the decimal place one place to the right.) Now our bonds currently sell for $965.00, so we must set up the percent problem:

$64 is what percent of $965?

Setting up the proportion, we have:

$$\frac{x}{100} = \frac{64}{965} \Rightarrow 965x = (100)(64) \Rightarrow x = 6,400/965 \Rightarrow x = 6.632$$

The current yield on PSvEG is 6.6%.

BROKERAGE FEES ON BONDS

Stockbrokers handle bonds as well as stocks. Fees vary from firm to firm, but the following table is fairly representative:

Brokerage Rates on Bonds
(Quotations per Bond)

Bond Quotation	Minimum Fee
30 but under 50	$2.75
50 but under 70	$4.25
70 but under 90	$6.75
Over 90	$8.50

EXAMPLE 4   Find the brokerage fee on 8 bonds sold at 69 ¾.

SOLUTION

The price is above 50 but below 70, so the brokerage fee is $4.25 per bond:

$$8 \times 4.25 = \$34.00$$

The brokerage fee will be $34.00 on these 8 bonds.

EXERCISES

Exercise A   Change each of the following bond price quotations to the actual price:

	Price Quotation	Price	
1.	$93\frac{1}{2}$	_____	__
2.	$85\frac{3}{8}$	_____	__
3.	$102\frac{1}{4}$	_____	__
4.	$47\frac{3}{4}$	_____	__
5.	$115\frac{1}{8}$	_____	__

	Price Quotation	Price	
6.	$94\frac{5}{8}$	_____	__
7.	$89\frac{1}{4}$	_____	__
8.	$95\frac{1}{8}$	_____	__

Convert each of the following net change quotations for bonds to the actual dollar equivalent:

	Net Change Quotation	Net Change	
9.	$+\dfrac{3}{4}$	_____	_____
10.	$-\dfrac{5}{8}$	_____	_____
11.	$-2\dfrac{1}{8}$	_____	_____
12.	$+1\dfrac{5}{8}$	_____	_____
13.	$-1\dfrac{3}{4}$	_____	_____

Refer to the bond market quotations on page 162 to find the interest rate and the maturity date for each of the following bonds:

	Company	Interest Rate	Maturity Year
14.	AMR	_____	_____
15.	TVA	_____	_____
16.	Deere	_____	_____
17.	Trump	_____	_____

Find the cost for the number of bonds purchased in each of the following transactions:

	Number of Bonds	Quotation per Bond	Price per Bond		Total Price	
18.	5	$68\frac{3}{4}$	————	——	————	——
19.	5	$75\frac{3}{4}$	————	——	————	——
20.	12	$67\frac{1}{2}$	————	——	————	——
21.	15	$58\frac{7}{8}$	————	——	————	——

Using the bond market quotation report on page 162, find the total cost for the number of bonds bought at the indicated quotations.

	Company	Number of Bonds	Price per Bond		Total Price	
22.	Amresco $\left(10s04\right)$	3	————	——	————	——
23.	Dole	12	————	——	————	——
24.	PepBoys	15	————	——	————	——
25.	SearsAc	13	————	——	————	——

Find the current yield on each of the following bonds:

	Quotation	Price Paid		Interest Rate	Annual Interest		Current Yield
26.	$102\frac{1}{4}$	————	——	$14\frac{1}{8}\%$	————	——	————
27.	$85\frac{3}{4}$	————	——	13	————	——	————

Exercise B   Find the total brokerage fee on each of the following bond sales:

	Number of Bonds	Market Value	Fee per Bond		Total Fee	
28.	5	$53\frac{1}{2}$	___	___	___	___
29.	8	$69\frac{1}{4}$	___	___	___	___
30.	15	$35\frac{7}{8}$	___	___	___	___
31.	6	$90\frac{1}{8}$	___	___	___	___
32.	14	$86\frac{1}{4}$	___	___	___	___

Find the total cost of each of the following bond purchases:

	Number of Bonds	Market Value	Fee per Bond		Total Fee		Total Cost	
33.	5	$95\frac{1}{2}$	___	___	___	___	___	___
34.	8	$69\frac{7}{8}$	___	___	___	___	___	___
35.	9	$86\frac{3}{4}$	___	___	___	___	___	___
36.	15	$48\frac{1}{8}$	___	___	___	___	___	___
37.	16	$65\frac{5}{8}$	___	___	___	___	___	___

Word Problems

Exercise C   Refer to the bond market report on page 162, and solve the following problems:

38. Mrs. Wilson bought 12 bonds of ATT, paying 6% annual interest and maturing in the year 2009.

   (a) How much did she pay for the bonds?
   (b) If she keeps the bonds for a full year, how much interest will she earn?
   (c) What is the current yield for each bond?

39. Mr. Romano bought 14 JCP bonds.

   (a) What was the cost of the 14 bonds?
   (b) How much was the brokerage fee for the sale?
   (c) What was the total cost of the bonds?

UNIT 5: Review of Chapter 5

Each of the following accounts had no withdrawals or deposits for the periods shown. Find the amount of interest and new balance for each account.

	Amount	Interest Rate	Period of Time	Amount of Interest		New Balance	
1.	$ 6,570	$6\frac{1}{4}\%$	3 years				
2.	8,340	13.263	30 months				
3.	15,600	12.725	6 months				
4.	4,400	14.72	3 months				

Using the blank forms provided, prepare deposit slips for the following deposits:

5.
DATE . 20	DOLLARS	CENTS
BILLS		
COINS		
CHECKS		
TOTAL		

6.
DATE . 20	DOLLARS	CENTS
BILLS		
COINS		
CHECKS		
TOTAL		

5. Bills: 7 $20's, 13 $10's, 16 $5's, 27 $1's
   Checks: $142.75, $193.68, $186.48, $137.64

6. Bills: 15 $20's, 23 $10's, 28 $5's, 29 $1's
   Checks: $143.79, $258.53, $178.38, $375.84

Using the blank form provided, prepare the following check. The balance brought forward is $816.92.

7.

No. _____

_____ 20 _____

Pay to the
order of _____ $ _____

_____ Dollars

For _____   _____

7. Check #143: To American Express for February statement; amount of check, $235.63; date of check, February 6; deposit, $423.75.

Prepare a reconciliation statement for the following account:

	Checkbook Balance	Bank Statement Balance	Outstanding Checks	Service Charge	Other Adjustment	Corrected Balance
8.	$ 880.93	$845.75	$ 89.45 115.67 142.25	$3.75	$378.80 deposit not on bank statement	_____

Rewrite each of the following stock prices in equivalent decimal forms:

9. $63 $\frac{1}{8}$ 10. $18 $\frac{7}{8}$ 11. $31 $\frac{5}{8}$

Find the net cost of the following lots of stocks at the indicated prices:

12. 225 at $62 $\frac{5}{8}$ 13. 650 at $8 $\frac{3}{8}$

Find the preceding day's closing price for each of the following stock prices:

14. $40 $\frac{1}{4}$ + 1$\frac{1}{8}$ 15. $47 $\frac{7}{8}$ + $\frac{5}{8}$

For Problems 16, refer to the stock market quotations on page 155. Find the price per share and the net price of 475 shares of the following stock that was bought at the price indicated:

	Company	When Bought	Price per Share	Net Total Price	
16.	JohnJn	Last	_____	_____	___

Find the brokerage fee on each of the following sales of stock:

	Number of Shares	Price per Share	Brokerage Fee	
17.	300	$23 $\frac{1}{8}$	_____	___
18.	75	53 $\frac{3}{8}$	_____	___

Find the total cost including brokerage fees for the following sale of stock:

	Number of Shares Bought	Price per Share	Selling Price		Brokerage Fee		Total Cost	
19.	600	$9\frac{7}{8}$	_____	__	__	__	_____	__

Find the net proceeds for each of the following sales:

	Number of Shares Sold	Price per Share	Selling Price		Brokerage Fee		Net Proceeds	
20.	300	$16\frac{1}{4}$	_____	__	__	__	_____	__
21.	100	$87\frac{7}{8}$	_____	__	__	__	_____	__

Change each of the following bond quotations to the actual price:

22. $105\frac{1}{8}$ 23. $95\frac{3}{4}$

Change each of the following net changes in bond quotations to the actual dollar equivalent:

24. $-1\frac{1}{4}$ 25. $+\frac{5}{8}$

Use the bond market quotations on page 162 to do the following problem:

	Company	Interest Rate	Maturity Year
26.	CentrTrst	_____	_____

Find the cost for the number of bonds purchased in each of the following transactions:

	Number of Bonds	Quotation per Bond	Price per Bond		Total Price	
27.	8	$78\frac{1}{4}$	_____	__	_____	__
28.	15	$61\frac{3}{8}$	_____	__	_____	__

Find the annual interest and current yield on the following bond:

	Quotation	Price Paid		Interest Rate	Annual Interest		Current Yield
29.	$87\frac{1}{8}$	___	___	$13\frac{1}{8}\%$	___	___	___

Find the brokerage fee on each of the following bond sales:

	Number of Bonds	Market Value	Fee per Bond		Total Fee	
30.	8	$79\frac{1}{4}$	___	___	___	___
31.	15	$88\frac{3}{4}$	___	___	___	___

Find the brokerage fee and total cost for the following purchase:

	Number of Bonds	Market Value	Fee per Bond		Total Fee		Total Cost	
32.	13	$99\frac{7}{8}$	___	___	___	___	___	___

Word Problems

Solve the following problems:

33. Maria has a savings account that earns an interest rate of 13.28%. If she earns a monthly interest of $134.50, how much money does she have in her savings account?

34. Joel needs $375 a month to supplement his education expenses. If a bank pays an interest rate of 13.7%, how much money will Joel have to deposit to earn the monthly amount that he needs?

35. Mark's checkbook shows a balance brought forward of $471.32. He wrote checks for $375.81, $167.25, and $87.77. In reviewing his check stubs, he noticed that he had made an error in entering a deposit of $386.95 as $586.95; he corrected this and also deposited a paycheck of $375.80. What should be his corrected balance carried forward on his last check stub?

Loans and Credit

Loans are how small businesses get enough money to carry on their business. Credit is the ability of a business to borrow money.

Unit 1: Promissory Notes

Banks make business and personal loans, but they are not the only ones who make loans. Individual investors, businesses, and credit unions also make loans. It is not uncommon for parents to make loans to their children to teach them the importance of borrowing carefully.

Sometimes these loans are informal and a handshake is all that is needed to conclude the deal. Most times, however, the terms of the loan are written down and signed by both parties. If the terms of the loan require interest to be paid, the note is called a **promissory note**. The borrower agrees to pay back the money and the agreed-to interest. For example, suppose that on August 23, 2000, Frank Cruz borrowed $450.00 from Hal Parker at 12% interest, to be paid back in 60 days. He prepared the following promissory note and gave it to Hal:

PROMISSORY NOTE	
$ _450 00/100_	_August 23_ 20_00_
60 days _____ after date _I_ _____ promise to pay to	
the order of _Hal Parker._	
Four hundred fifty and 00/100 _____Dollars	
at _123 Main Road, Anytown., N.Y._	
Value received _12%_	
No. _31_ _____Due _October 22, 2000_	
76 121	_Frank Cruz_

In this note,

- $450.00 is the **face value** or **principal** of the promissory note.
- August 23, 2000 is the **date** of the note.
- 60 days are the **terms**.
- Hal Parker is the **payee**.
- Frank Cruz is the **maker**.
- October 22, 2000 is the **due date** or the **date of maturity**.
- 12% is the yearly **rate of interest**.

This last term needs some explanation. 12% is the *yearly* rate of interest. However, the note is for only 60 days. Therefore, the interest will be calculated for 60/360, or 1/6, of the full 12% yearly interest. We will calculate this in one of the examples.

Find the maturity date on a 60-day promissory note dated July 9.

SOLUTION

Two months after July is September. July has 31 days, so the maturity date of the note is September 8.

Find the maturity date on a 90-day promissory note dated November 20.

SOLUTION

This is a bit tricky because we must go into the next year. Three months past November is February, but December has 31 days, so the maturity date is one day earlier, February 19.

Hint: To remember the number of days in each month, try to memorize this little ditty:

> "Thirty days has September, April, June and November, all the rest have thirty-one, except February having 28, and 29 in a leap year."

Another way of remembering the months that have 30 and 31 days is to make a fist and look at your knuckles. The knuckles represent the months that have 31 days, being high. The valleys between the knuckles represent the months that have 30 days, being low. The exception is February, which has 28 days and has 29 days in a leap year.

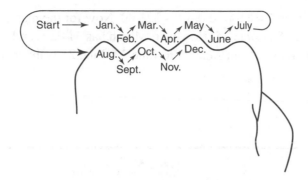

FINDING THE MATURITY VALUE

 Find the maturity value on the promissory note shown at the beginning of this unit.

SOLUTION

Remember that the promissory note was for 60 days (the terms), the rate of interest was 12% per annum, and the principal was $450.00. The equation is set up this way:

$$I = P \times R \times T$$
$$I = 450 \times 0.12 \times 60/360$$

Note that percent is written as a decimal. The time is 60/360 because the note is for 60 days. Also, recall that most businesses use 360 instead of 365 days for a year.

$$I = 450 \times 0.12 \times 0.1666$$
$$I = 9$$

The interest on $450 at 12% for 60 days is $9.00.

 Find the maturity value of a $600 note at 14% for 90 days and that is dated March 25. On what date does this note mature?

SOLUTION

We have principal: $600; rate of interest: 14%; terms: 90 days. The equation is set up this way:

$$I = P \times R \times T$$
$$I = 600 \times 0.14 \times 90/360$$

Note that percent is written as a decimal. The time is 90/360 because the note is for 90 days.

$$I = 600 \times 0.14 \times 0.25$$
$$I = 21$$

The interest on $600 at 14% for 90 days is $21.00. Now for the maturity date. Three months from March 25 is June 25. However, March has 31 days and so does May. So we must back up two days from June 25 to June 23. June 23 is the maturity date of the promissory note.

EXERCISES

Exercise A   Find the maturity date for each of the following promissory notes:

	Date of Note	Terms	Maturity Date
1.	September 19	3 months	_____
2.	August 12	20 days	_____
3.	January 21	30 days	_____
4.	June 16	8 months	_____
5.	July 10	60 days	_____
6.	March 9	45 days	_____
7.	June 16	90 days	_____
8.	December 18	2 months	_____
9.	August 28	6 months	_____
10.	May 20	90 days	_____

Find the maturity date, interest, and maturity value of each of the following promissory notes:

	Face Value	Date of Note	Rate	Terms	Due Date	Interest		Maturity Value	
11.	$ 790	July 3	13%	90 days	_____	___	___	___	___
12.	475	Nov. 12	8.5	3 years	_____	___	___	___	___
13.	950	Oct. 1	12	120 days	_____	___	___	___	___
14.	565	June 8	$10\frac{1}{2}$	40 days	_____	___	___	___	___
15.	765	Mar. 12	$9\frac{3}{4}$	90 days	_____	___	___	___	___

Word Problems

Exercise B   Solve the following problems:

16. Find the due date and maturity value of a promissory note for $675 dated March 9, for 90 days at 14% interest.

17. On July 11 Jim borrowed $1,575 at 12% interest for 75 days and signed a promissory note. When will the note become due, and how much will he have to pay back?

18. Find the face value of a note for 1 year at 9% if the amount of interest is $108.

Unit 2: Borrowing from a Bank

Banks lend money to people for all sorts of reasons: for college tuition, to buy a car, to buy a house, for repairs around the house, and many, many other needs. Banks make loans to individuals based on their **credit standing**, which is calculated based on many factors. You might consider finding out what goes into a credit report. One of the many web sites that explains how your credit standing is calculated is *creditreport. com*. Sometimes a bank will require a **cosigner**, another individual who will guarantee the repaying of the loan. Very often this is a parent in the case of a student taking out a student loan. Sometimes a bank will require **collateral**, a temporary transfer of a possession to the bank until the loan has been paid off.

A loan agreement from a bank is much like the promissory note of the previous unit. It generally is a multipage document, not a single slip of paper.

CALCULATING INTEREST

To calculate interest, we use the same formula we have always used:

$$I = P \times R \times T$$

 EXAMPLE 1   Find the interest on a loan of $2,500 at 11% for 3 years.

SOLUTION

The equation is set up this way:

$$
\begin{aligned}
I &= P \times R \times T \\
&= 2{,}500 \times 0.11 \times 3 \\
&= 825
\end{aligned}
$$

The interest on a loan of $2,500 at 11% for 3 years is $825.00.

Find the interest on a loan of $1,250 at 8% for 180 days.

SOLUTION

The equation is set up this way:

$$I = P \times R \times T$$
$$= 1{,}250 \times 0.08 \times 180/360$$
$$= 1{,}250 \times 0.08 \times .5$$
$$= 50$$

The interest on a loan of $1,250 at 8% for 180 days is $50.00.

NET AMOUNT AND MONTHLY PAYMENTS OF LOANS

On many personal loans, a bank will **discount** the loan. That is, the bank will calculate the interest, subtract that amount from the principal, and give that amount to the borrower. That is called the **net amount**. The borrower will then pay that amount back in monthly payments for the duration of the loan.

Marie borrows $5,370 from a bank at 9% for 3 years. The interest is discounted, and she must pay the loan back in 36 monthly payments. Find the net amount that Marie will receive and the amount of her monthly payments.

SOLUTION

The equation is set up this way:

$$I = P \times R \times T$$
$$= 5{,}370 \times 0.09 \times 3$$
$$= 1{,}449.9$$

The interest on a loan of $5,370 at 9% for 3 years is $1,449.90.

Next, we calculate the net amount:

$$\begin{array}{r} 5{,}370.00 \\ -\ 1{,}449.90 \\ \hline 3{,}920.10 \end{array}$$

Marie will get $3,920.10 from the bank. Now we must divide this into 36 equal payments:

$$3{,}920.10/36 = 108.89$$

Marie's monthly payments will be $108.89.

THE 12% 30-DAY METHOD

Bankers, like most people, have little shortcuts, or rules of thumb, to calculate payments easier. For short-term loans, it is sometimes convenient to use the 12% 30-day rule of calculating interest. The rule is this:

> *One month (30 days) is 1/12 of a 360-day (commercial) year. If the interest rate is 12% per year, then the rate for 30 days is 1/12 of 12%, or 1%.*

This rule can be used for other periods of days or months.

 Find the interest on a loan of $925 at 12% interest for 30 days.

SOLUTION

This uses the 12% 30-day rule exactly. Multiply:

$$925 \times 0.01 = 9.25$$

So interest on a loan of $925 at 12% for 30 days is $9.25.

 Find the interest on a loan of $1,150 at 12% interest for 30 days.

SOLUTION

This uses the 12% 30-day rule exactly. Multiply:

$$1,150 \times 0.01 = 11.50$$

So interest on a loan of $1,150 at 12% for 30 days is $11.50.

USING PARTS AND COMBINATIONS OF THE 12% 30-DAY RULE

We can make adjustments to the 12% 30-day rule rather easily. For instance, if the length of the loan is 15 days, the interest would be 0.5% (half of 1%).

 Find the interest on a loan of $4,150 at 12% interest for 45 days.

SOLUTION

This uses the 12% 30-day rule 1½ times. Multiply:

$$4,150 \times 0.01 = 41.50$$

This is for 30 days. However, we have 45 days, so we take half of 41.50:

$$41.50/2 = 20.75$$

then add that to the 41.50:

$$41.50 + 20.75 = 62.25$$

So on a loan of \$4,150.00 for 45 days, the interest will be \$62.25.

 EXAMPLE 7   Find the interest on a loan of \$3,550 at 12% interest for 90 days.

SOLUTION

This uses the 12% 30-day rule 3 times. Multiply:

$$3,550 \times 0.03 = 106.50$$

So on a loan of \$3,550.00 for 90 days, the interest will be \$106.50.

The following table lists the most commonly used parts and combinations of the 12% 30-day rule:

Using Parts and Combinations of 30 Days

Parts	Combinations
5 days = $\frac{1}{6}$	35 days = $1\frac{1}{6}$
6 days = $\frac{1}{5}$	40 days = $1\frac{1}{3}$
10 days = $\frac{1}{3}$	45 days = $1\frac{1}{2}$
15 days = $\frac{1}{2}$	60 days = 2
20 days = $\frac{2}{3}$	90 days = 3
25 days = $\frac{5}{6}$	120 days = 4
	150 days = 5
	180 days = 6
	210 days = 7
	240 days = 8
	300 days = 10

EXERCISES

Exercise A   Find the interest and the net amount on each of the following discounted loans:

	Principal	Rate	Time	Amount of Interest		Net Amount	
1.	$1,285	10%	2 years	———	—	———	—
2.	2,350	13	60 days	———	—	———	—
3.	5,425	$10\frac{1}{2}$	4 years	———	—	———	—
4.	2,875	8.75	$1\frac{1}{2}$ years	———	—	———	—
5.	2,575	$9\frac{1}{4}$	9 months	———	—	———	—

Find the interest, the net amount, and the monthly payment on each of the following discounted loans:

	Principal	Rate	Time	Amount of Interest		Net Amount		Monthly Payment	
6.	$1,875	13%	18 months	———	—	———	—	———	—
7.	4,560	$11\frac{1}{2}$	36 months	———	—	———	—	———	—
8.	2,750	13	24 months	———	—	———	—	———	—
9.	2,575	15	18 months	———	—	———	—	———	—
10.	4,500	11.75	48 months	———	—	———	—	———	—

Without using written calculations, find the interest at 12% on the following loans:

11. $685 for 60 days   12. $2,400 for 30 days   13. $5,260.75 for 60 days

14. $3,237.80 for 60 days   15. $5,000 for 240 days

Find the interest and the total amount due on each of the following loans at 12% interest:

	Principal	Time	Interest		Amount Due	
16.	$3,475	60 days	_____	____	_____	____
17.	2,463	150 days	_____	____	_____	____
18.	4,295	10 days	_____	____	_____	____
19.	2,560	75 days	_____	____	_____	____
20.	4,290	90 days	_____	____	_____	____

Word Problems

Exercise B Solve the following problems:

21. Sandy plans to borrow $1,250 to pay for a vacation. The bank charges 16% interest. Sandy expects to repay the loan in 18 months. If the bank discounts the interest, what is the net amount she will receive from the bank?

22. John needs $1,800 to pay his college tuition. If the bank charges 11% and he plans to pay off the loan in 1 year, how much should the amount of the loan be?

23. Jim took out a loan of $4,320 at 13% for 4 years. What was the net amount of the loan?

24. What is the total amount due the bank on a $1,370 loan at 12% for 45 days?

25. Find the interest on a $1,765 loan at 12% for 40 days.

Unit 3: Credit Buying

Credit buying is a very common practice in America. Most department stores, car companies, and many other kinds of companies issue credit cards. Banks also offer credit cards to their customers. There are two main kinds of credit purchasing: a **credit card** and **installment buying**.

A credit card is offered to individuals who have undergone a credit check. The person's financial background has been scrutinized, and the individual has been deemed credit worthy. The credit card can then be used to buy items in any store that accepts the card up to the credit limit on the card. At the end of the month, paying off the card generally results in no interest payment. If the balance is not paid off, there is an interest charge on the unpaid balance.

INSTALLMENT BUYING

Large purchases, such as furniture, home appliances, automobiles, and power tools, can be paid for over time in an installment plan. Instead of paying cash or using a credit card, the buyer pays a **down payment** and is then charged a certain sum every month until the balance is paid off. These are usually equal monthly installments. The total price of an item is usually higher than the full cash price. The difference between the full price and the installment price is called the **carrying charge**, and it is much like interest on the purchase. The carrying charge is somewhat higher than an interest-bearing credit card purchase, because it also includes the clerical costs the store has to service the installments.

CALCULATING CARRYING CHARGES

The cost of a riding mower is $894.00. The retailer sells it for $150 down and 12 installment payments of $70.00. What is the total cost of the mower? What is the carrying charge?

SOLUTION

First we find out what the total cost of the mower is:

Down payment:	$150
Monthly payments:	$70 × 12 = $840
Total cost:	$150 + $840 = $990.00
Carrying charge:	Total cost − Retail price
	$990 − $894 = $96.00

The total cost is $990.00.
The carrying charge is $96.00.

Some retailers will display these charges on the merchandise, including the down payment, the carrying charges, and the number of monthly payments. Some will disclose them only if asked. Be sure you know what financial obligations you are committing to before you make any large purchase.

The total cost (installment payment cost) of a dining room set is $2650.00. The down payment is $250, and the buyer will pay the balance in 48 equal installments. How much will each installment be?

SOLUTION

First we find out what the balance is:

Total cost − Down Payment = Balance
$2,650 − $250 = $2,400.00

Then we calculate the monthly installments:

$$\$2,400/48 = 50$$

The monthly installments on this dining room set will be $50.00 per month for 48 months.

CALCULATING THE TRUE INTEREST RATE

One of the advantages of installment buying is that you can immediately have appliances and other large items that you cannot otherwise afford. The disadvantage of this system is that interest and carrying charges can be expensive and, in fact, can drown some people in debt.

There was a time when the true cost of an item was difficult to find and calculate. Since the passage of the Truth-in-Lending Law, rates have been much more transparent.

The formula for the true interest rate (TIR) is:

$$R = \frac{2YC}{B(N + 1)}$$

Where: R = true rate of interest

Y = 12 months (1 year)

C = carrying charge

B = balance of the cash price (minus the down payment)

N = number of installments

 EXAMPLE 3 Roger bought a high-definition flat-screen TV for $750 cash, or $150 down and 12 installments of $58. What is the true interest rate?

SOLUTION

First we find the total cost (installment price):

$$150 + 12 \times 58 = 150 + 696 = 846$$

The total cost is $846.00. Next we calculate the carrying charge:

$$\$846 - \$750 = \$96$$

Finally, we substitute these figures into the true interest rate equation:

$$R = \frac{2 \times 12 \times 96}{600(12 + 1)} = \frac{2,304}{7,800} = 0.295$$

The true interest rate is 29.5%.

EXERCISES

Exercise A   Find the carrying charge for each of the following installment purchases:

	Item	Cash Price	Down Payment	No. of Payments	Monthly Payments	Carrying Charge	
1.	Carpeting	$ 775	$110	24	$33	_____	_____
2.	Washer	245	65	18	12	_____	_____
3.	Piano	1,250	225	36	45	_____	_____
4.	Diamond ring	525	85	24	27	_____	_____
5.	Dishwasher	185	45	18	13	_____	_____

Find the amount of each payment for the following installment purchases:

	Item	Installment Price		Down Payment	Number of Payments	Amount of Each Payment	
6.	A	$248	95	$53	12	_____	_____
7	B	428	65	85	36	_____	_____
8.	C	753	60	115	36	_____	_____
9.	D	438	65	85	18	_____	_____
10.	E	463	85	82	18	_____	_____

Find the true rate of interest on each of the following installment purchases:

	Item	Cash Price	Down Payment	Number of Payments	Monthly Payments	True Rate of Interest
11.	F	$750	$130	24	$35	_____
12.	G	255	75	18	12	_____
13.	H	1,375	250	36	48	_____
14.	I	650	110	24	30	_____
15.	J	510	75	24	26	_____

Word Problems

Exercise B   Solve the following problems:

16. A TV set can be bought for $465 cash or on the installment plan with a $50 down payment and 12 payments of $38 each. Find the carrying charges if the set is purchased on the installment plan.

17. The installment price of a freezer is $685. If the down payment is $75 and the balance is to be paid in 24 equal installments, how much will each installment be?

18. Alfredo bought a radio for $138.95. He made a down payment of $35 and will pay off the balance in 1 year. If the finance charge is $18.65, what will be the total balance to be paid?

Unit 4: Review of Chapter 6

HINTS: • "30 days have September, April, June, and November. All the rest have 31, except February, which has 28, and in Leap Year, 29."

• The 12%-30 days method: 12% interest for 1 year is equivalent to 1% interest for 30 days.

Find the maturity date for each of the following promissory notes:

	Date of Note	Terms	Maturity Date
1.	February 15	4 months	_____
2.	April 20	90 days	_____
3.	December 18	45 days	_____

Find the maturity date, interest, and maturity value of each of the following promissory notes:

	Face Value	Date of Note	Rate	Terms	Due Date	Interest	Maturity Value
4.	$2,575	Aug. 25	13%	3 months	_____	_____ _____	_____ _____
5.	1,500	May 21	$15\frac{1}{4}$	45 days	_____	_____ _____	_____ _____

Find the interest and the total debt on each of the following loans:

	Principal	Rate	Time	Interest		Total Debt	
6.	$1,350	12.25%	3 years	_____	___	_____	___
7.	3,875	14.5	18 months	_____	___	_____	___
8.	4,125	$14\frac{3}{4}$	$1\frac{1}{2}$ years	_____	___	_____	___

Find the interest, net amount, and monthly payment on each of the following loans:

	Principal	Rate	Time	Interest		Net Amount		Monthly Payment	
9.	$5,690	13.25%	24 months	_____	___	_____	___	_____	___
10.	6,540	13	48 months	_____	___	_____	___	_____	___

Find the carrying charge for the following installment purchase:

	Item	Cash Price	Down Payment	Number of Payments	Monthly Payments	Carrying Charge	
11.	A	$1,475	$245	36	$45	_____	___

Find the amount of each payment on each of the following installment purchases:

	Item	Installment Price		Down Payment	Number of Payments	Amount of Each Payment	
12.	1	$1,342	75	$325	36	___	___
13.	2	1,168	85	285	18	___	___

Home, Car, and Insurance Expenditures

Unit 1: Home Ownership

Buying a home is the largest purchase most families in America will make. It would be out of reach of most everyone if it were not for a **mortgage**, which is a loan from a bank or other lending institution for the price of the house. Home ownership is one of the most powerful threads in the fabric of the society. Owning a home is a viable option to many Americans because it allows them to build up equity in property that belongs to them. In contrast, for those in apartments, their dwelling belongs to someone else. Over time, the rent in an apartment can approach a mortgage payment but with no buildup of equity, as there is in a home.

Nonetheless, home ownership involves a number of expenses. In addition to the down payment and the mortgage payment, there are property taxes, home insurance, utilities, and repairs to the home. There are several kinds of mortgages. We will discuss two kinds: **fixed rate** and **variable rate**.

COMPUTING MORTGAGE PAYMENTS

The buyer of a house makes a down payment (usually 10–20% of the cost of the house), and then borrows the balance from a bank or other lending institution. The buyer then signs a **mortgage agreement**, promising to pay monthly payments according to the schedule calculated and set down in the mortgage agreement. Mortgages are usually made for 15 to 30 years. The interest rate can vary from as low as 5% up to 15%. Mortgage interest rates have gone as high as 19%, although that is very rare. In the case of a mortgage, the collateral for the agreement is the house itself. The bank or other lending institution could take possession of the house if the buyer fails to make the mortgage payments.

To find the amount of the monthly payments, an **amortization table** is consulted. An amortization table shows the amount of money needed to be paid on a unit of the mortgage (usually $1,000). It shows also the payments on multiple interest rates and multiyear loans. Here is a small portion of an amortization table:

TABLE 1
MONTHLY PAYMENT SCHEDULE FOR EACH $1,000 OF A MORTGAGE

Duration of Mortgage (years)	Interest Rate (%)							
	10	$10\frac{1}{2}$	11	$11\frac{1}{2}$	12	$12\frac{1}{2}$	13	$13\frac{1}{2}$
15	$10.75	$11.06	$11.37	$11.69	$12.01	$12.33	$12.66	$12.99
20	9.66	9.99	10.33	10.67	11.02	11.37	11.72	12.08
25	9.09	9.45	9.81	10.17	10.54	10.91	11.28	11.66
30	8.78	9.15	9.53	9.91	10.29	10.68	11.07	11.46

How to use this table:

1. Find out how many thousands of dollars the mortgage is by dividing the mortgage by 1,000.

2. Choose the number of years the mortgage is set for (15, 20, 25, or 30) and go down the left column to that time.

3. Choose the interest rate the mortgage is set for across the top row.

4. Find the monthly payment at the intersection of the years and the interest rate (found in steps 2 and 3).

5. Multiply the number from step 1 by the number found in step 4.

Mr. and Mrs. Maturin are buying a house in Dumont. The cost of the house is $175,500. They were able to secure a 30-year fixed mortgage at a rate of 10.5%. Calculate their monthly payments.

SOLUTION

First calculate the thousands of dollars the mortgage is:

$$175,500/1,000 = 175.5$$

Second, look down the column to the 30-year entry.
Third, look across the row to the 10.5% entry.
Fourth, find the monthly payment per thousand—9.15.
Fifth, multiply the number found in step 1 by the number found in step 4:

$$175.5 \times 9.15 = 1,605.825$$

Mr. and Mrs. Maturin's monthly payments will be $1,605.83.

Jade is buying a house in Callan. The cost of the house is $86,250. She negotiated a 20-year fixed mortgage at a rate of 11.5%. Calculate her monthly payments.

SOLUTION

First calculate the thousands of dollars the mortgage is:

$$86,250/1,000 = 86.25$$

Second, look down the column to the 20-year entry.
Third, look across the row to the 11.5% entry.
Fourth, find the monthly payment per thousand—10.67.
Fifth, multiply the number found in step 1 by the number found in step 4:

$$86.25 \times 10.67 = 920.2875$$

Jude's monthly payments will be $920.29

In these examples, the answer gives only the monthly payments, which are the principal and interest (for 20 or 30 years), without other expenses. Now we will look at how to calculate taxes and insurance.

PROPERTY TAXES

Property taxes are also called **real estate taxes**. They are a major contributor to a town's or city's budget to pay for police, fire, and schools, among other things. Property taxes are assessed on houses, buildings, and other property in the town or city. The amount of tax is based on the **assessed value** of the property. The assessed value is, in turn, based on the **market value**. The market value is an estimate of what the property would sell for if it was sold today. Many factors are used to calculate the market value (square footage, the property's condition, surrounding properties values, and so on). Since the market value fluctuates from year to year, the assessed value is recalculated once every several years for consistency.

Towns and cities calculate the property taxes differently. Generally, property taxes are a percentage of the assessed value of the property, which is a percentage of the market value. The tax rate is usually expressed as a dollar amount per thousand or per hundred, as in the following examples:

The assessed value of the Petri house is $115,500. The property tax rate in their town is $35.53 per thousand. How much will they pay in property tax?

SOLUTION

First divide the assessed value by 1,000:

$$115,500/1,000 = 115.5$$

Now multiply this number by the property tax rate:

$$115.5 \times 35.53 = 4,103.715$$

The property tax on the Petri house is $4,103.72.

EXAMPLE 4

Mr. Valentine's condominium is assessed at $75,350. The property tax rate in his town is $4.795 per hundred. How much will he pay in property tax?

SOLUTION

First divide the assessed value by 100:

$$75,350/100 = 753.5$$

Now multiply this number by the property tax rate:

$$753.5 \times 4.795 = 3,613.0325$$

The property tax on Mr. Valentine's condo is $3,613.03.

Some municipalities tax their properties differently. They will calculate the tax in **mills** per $1. Mills, are like the metric prefix for "thousand." It means thousandths of a dollar. To calculate when given mills, divide the number by 1,000 and then multiply by the assessed value. Other municipalities calculate tax in cents per $1. This is similar, but the cents are divided by 100. Then that number is multiplied by the assessed value. The following examples illustrate these.

EXAMPLE 5

Gene has a co-op apartment that is assessed at $65,585. The property tax in his town is 48 mills per $1. How much is his property tax?

SOLUTION

First divide mills by 1,000:

$$48/1,000 = 0.048$$

Now multiply this number by the assessed value:

$$65,585 \times 0.048 = 3,148.08$$

The property tax on Gene's co-op is $3,148.08.

EXAMPLE 6

Opie owns a house assessed at $285,350. The property tax in his town is 7.35 cents per $1. How much will he pay in property tax?

SOLUTION

First divide cents by 100:

$$7.35/100 = 0.0735$$

Now multiply this number by the assessed value:

$$285,350 \times 0.0735 = 20,973.225$$

The property tax on Opie's house will be $20,973.23

EXERCISES

Exercise A   Using the monthly payment schedule (Table 1) on page 190, find the monthly and yearly payments for each of the following mortgages:

	Mortgage	Rate	Period of Years	Monthly Payments		Yearly Payments	
1.	$45,350	$12\frac{1}{2}$ %	30	_____	___	_____	___
2.	55,675	$10\frac{1}{2}$	25	_____	___	_____	___
3.	64,500	$13\frac{1}{2}$	25	_____	___	_____	___
4.	51,570	$10\frac{1}{2}$	25	_____	___	_____	___
5.	32,410	$11\frac{1}{2}$	25	_____	___	_____	___

Find the real estate tax on each of the following properties:

	Assessed Valuation	Tax Rate	Amount of Tax	
6.	$12,500	$7.45 per $100	_____	___
7.	16,000	$62.35 per $1,000	_____	___
8.	21,500	7.23 cents per $1	_____	___
9.	17,500	$6.28 per $1,000	_____	___
10.	9,500	$6.43 per $100	_____	___

Word Problems

Exercise B   Use the monthly payment schedule (Table 1) on page 190 in solving the following problems:

11. Mr. Carlsen bought a $78,750 house. He made a 15% down payment and was granted a 30-year mortgage at $13\frac{1}{2}$% for the remainder. Find his monthly payments.

12. A house sells for $93,750. If the purchaser is required to make a 20% down payment and pay the remainder with a 30-year mortgage at $11\frac{1}{2}$%, how much will the monthly payments be?

13. A real estate agent offers a house for $48,950. If this includes a 15% commission for the agent, what would be the price of the house if bought directly from the seller?

14. Mr. Bondi owns two pieces of property assessed at $9,500 and $7,250. What will be his tax for this year if the rate is 7.15% of the assessed value?

15. If the tax rate is 53 mills per $1, how much will the tax be on property assessed at $19,650?

Unit 2: Home Insurance

If your house burns down, is damaged by a car, is robbed, or experiences some other mishap, how will you repair or replace it? Insurance companies will insure your home against a host of calamities. You must, though, read the policy carefully to know what, exactly, the insurance will cover. The company selling the insurance is called the **insurer**, and the homeowner purchasing the insurance is called the **insured**. The contract between the insurer and the insured is called an **insurance policy**. The amount of insurance covered by the policy is called the **face** of the policy.

The basic insurance policy on a building is for fire insurance. Other losses can be added to an insurance policy, like water and smoke damage from putting out a fire. However, coverage for every extra loss will raise the insurance **premium**, the price you pay for the insurance. An insurance policy that has more than the basic fire insurance is called an **extended-coverage** policy. Premium rates are generally based on $100 units of insurance for 1 year. Insurance companies rate buildings based on a number of factors such as where the building is located, what it is used for, the construction of the building, and so on. The following table shows a typical set of **term rates**. As you can see, insurance companies try to encourage homeowners to lengthen the time of their policy by offering lower rates per year if they take out multiple-year contracts:

TABLE 2
TERM RATES

Period	Term Rate
2 years	1.85 times annual rate
3 years	2.7 times annual rate
4 years	3.55 times annual rate
5 years	4.4 times annual rate

If a policy has an annual rate of $0.26 per $100, a 3-year policy would be 2.7 ×
$0.26, or $0.702, per $100. In other words, the policy would cost $0.23 per $100
per year (0.702 ÷ 3 = 0.234), a savings of 3 cents per $100 per year.

Jack's house is insured for $78,950. If the rate is $0.31 per $100, what will be the
insurance premium?

SOLUTION

First find the number of $100 units:

$$78,950/100 = 789.5$$

Now multiply the result by 0.31 to find the premium:

$$789.5 \times 0.31 = 244.745$$

The premium on this policy will be $244.75.

Jill has a 5-year plan with her insurance company. The house is insured for
$65,435. If the annual rate is $0.27 per $100, calculate the premium. Use the infor-
mation in Table 2 to calculate the premium.

SOLUTION

First find the number of $100 units:

$$65,435/100 = 654.35$$

Now look at Table 2. For a 5-year policy, we multiply by 4.4:

$$0.27 \times 4.4 \times 1.188$$

Now multiply 654.35 by 1.188 to find the premium:

$$654.35 \times 1.188 = \$777.3678$$

The premium on this policy will be $777.37.

CANCELING A POLICY

An insurance policy is taken out for a year or more. The homeowner rarely, however,
sells his/her house on the exact day that the policy expires. That is why there is a formula for refunding the price of the unused portion of the policy. This is called a **pro
rata basis**. It is a fairly straightforward formula. The number of days the policy was in
effect is divided by the year, or two, that the policy was contracted for, and that fraction is multiplied by the premium to get the prorated premium.

EXAMPLE 3

An insurance plan on a building has a premium of $647.54 for two years. The insured sells the building, however, after 405 days. How much of the premium was refunded to the owner?

SOLUTION

First subtract to get the number of days the policy is not in effect.

$$730 - 405 = 325$$

Remember, the policy is for 2 years—730 days: Now divide the number of days the policy will not be in effect by the number of days the policy was originally set for:

$$325/730 = 0.4452$$

Now multiply this number by the premium:

$$647.54 \times 0.4452 = 288.288$$

The refund on the premium will be $288.29.

EXERCISES

Exercise A   Find the premium for 1 year on each of the following policies:

	Face Value of Policy	Annual Rate per $100	Annual Premium	
1.	$68,500	$ 0.35	_____	___
2.	61,750	0.29	_____	___
3.	54,000	0.36	_____	___

Find the total premium on each of the following policies for the terms indicated:

	Face Value of Policy	Annual Rate per $100	Term of Policy	Total Premium	
4.	$69,750	$0.36	4 years	_____	___
5.	68,750	0.32	5 years	_____	___

Find the unexpired days and the amount of refund on each of the following canceled yearly policies.

	Annual Premium	Policy in Force	Unexpired Days	Amount of Refund
6.	$186	75 days	_____	_____
7.	224	235 days	_____	_____
8.	215	180 days	_____	_____

Word Problems

Exercise B   Solve the following problems:

9. Jennifer Lodge insured her house, which is valued at $53,000, for 3 years. The annual rate is $0.55 per $100. If the 3-year term rate is 2.7 times the annual rate, how much is the total premium for the 3-year policy?

10. Ben Singer wants to insure his home, which is valued at $63,000. The annual rate is $0.56 per $100, and the term rate for 5 years is 4.4 times the annual rate. How much would Mr. Singer save by insuring his home under the 5-year term rate?

Unit 3: Automobile Ownership

If home ownership is the largest purchase a family will make, the second-largest purchase will be an automobile. Two major costs are associated with an automobile. The first is the purchase price, including the down payment and monthly payments. The second expense is the operating costs, including the insurance and fuel costs, the parking costs, and the repair and maintenance costs.

An automobile loses its value over time, and this is called the **depreciation** cost. A car gets worn down and more prone to breakdowns the older it gets. This loss in value is what is meant when the depreciation is calculated. If you buy a car for $19,000 and it loses $11,000 worth of its original value in 6 years, then it is worth $8,000.

COMPUTING THE ANNUAL DEPRECIATION

The greatest depreciation on a car occurs during the first year. Indeed, driving a new car off the dealer's lot depreciates the car $1,000 to $2,000. Depreciation decreases in succeeding years. It is best, therefore, if the car owner keeps the car for as long as possible.

There are formulas for calculating depreciation, but they are complicated. A publication, the *Blue Book*, tells what the value of any make and model of car is after a certain number of years of ownership. We will not consider those methods. We will use a straight-line method, where the car loses its value in a linear (straight-line) fashion over the useful life of the car.

EXAMPLE 1   Tracy bought a car for $22,500.00. After 6 years, the car was worth $12,000.00. What was the average annual depreciation?

SOLUTION

First calculate the total depreciation:

$$\$22,500 - \$12,000 = \$10,500$$

Now calculate the average annual depreciation:

$$10,500/6 = 1,750$$

The average annual depreciation of the car was $1,750.00.

EXAMPLE 2   Jay and Betsy bought a car for $18,650.00. The average annual depreciation was $1,245.00. What was rate of depreciation?

SOLUTION

First divide the average annual depreciation by the original cost of the car:

$$1,245/18,650 = 0.066756$$

Now change the decimal into a percent:

$$0.066756 \times 100 = 6.7\%$$

The rate of depreciation of the car was 6.7%.

EXERCISES

Exercise A   Find the annual depreciation in each of the following:

	Original Cost	Trade-In Value		Annual Depreciation
		At the End of	Amount	
1.	$6,800	4 years	$1,300	_____ __
2.	9,800	6 years	1,100	_____ __
3.	8,650	5 years	900	_____ __

Find the annual depreciation and the rate of depreciation to the nearest percent in each of the following:

	Original Cost	Trade-In Value		Annual Depreciation	Rate of Depreciation
		At the End of	Amount		
4.	$9,600	4 years	$1,500	_____	____
5.	7,800	6 years	900	_____	____

Word Problems

Exercise B   Solve the following problems:

6. Todd bought a car for $9,350. Six years later he sold it for $1,150. How much was the annual depreciation of the car?

7. Floyd bought a car for $9,500. Five years later he was allowed $1,150 for it toward the purchase of a new car. Find, to the nearest percent, the annual rate of depreciation.

8. After keeping his car for 4 years, Marty sold it for $2,150. If he originally paid $7,275, what was the annual depreciation of the car?

Unit 4: Automobile Insurance

Insuring a car is even more necessary than homeowner's insurance, because automobile accidents happen much more often than home mishaps. There are four different components to a full automobile insurance policy: bodily injury, property damage, collision, and comprehensive insurance.

Bodily injury insurance protects the car owner against financial loss due to the killing or serious injury of a person by the car. Most states require this insurance in order for the car to be registered and licensed in the state. A typical bodily injury policy would be a 25/50 policy. This means that in the event of an accident, the insurance company will pay any one person up to $25,000 injured and up to $50,000 to all people injured in an accident. As an example, suppose you were in an automobile accident and three people were injured. (Whether on not they were in your car does not matter.) The court awarded one person $28,000 and the other two people $5,500 and $7,500. Your insurance company would pay $25,000 to that first person and all the money to the second and third persons. You, however, would need to pay the first person another $3,000 because the insurance company will award only the maximum amount of the policy, not the amount the person was awarded by the court. Higher amounts are available. Bodily injury coverage of 50/100, 100/300, and even higher are not uncommon.

Property damage insurance protects the car owner against any damage to other people's property, such as another car, a house, a building, a storefront, and so on. Coverage for property damage is generally in the $50,000 to $100,000 range.

Collision insurance will pay for damage done to your car as a result of an accident, whether it be with another car, a house or other object (a fire hydrant, for instance), a blowout, or any other mishap. Collision insurance usually has a deductible amount, which makes the cost of the insurance lower. A deductible amount means that if the damage is less than the deductible amount, you would need to pay for the damage. A typical deductible amount is $100 to $500.

Comprehensive insurance is for loss of your car due to factors other than an accident, such as theft, vandalism, fire, or flood. There usually is a deductible amount associated with this insurance as well.

For both collision and comprehensive insurance, the insurance company would pay up to the *Blue Book* value of the car in the event it is a total loss.

Insurance premiums for bodily injury and property damage are set based on four factors:

1. The frequency of accidents in the area the insured lives

2. The age of the principal driver

3. The principal use of the car

4. The number of accidents the principal driver has had in the last 3 years

Every state is divided into territories. The base premium is set higher in territories that have a higher frequency of accidents. Often, this means that a rural driver will pay less in insurance that an urban driver because more accidents occur in a city.

The following is a sample base premium table:

TABLE 3
BASE PREMIUMS FOR PRIVATE PASSENGER AUTOMOBILES

Type and Amount of Policy	Territory					
	01	02	03	04	05	06
Bodily Injury						
$ 10,000/$20,000	$84	$76	$63	$49	$32	$28
25,000/50,000	86	77	66	52	35	31
50,000/100,000	93	81	70	56	39	35
100,000/300,000	98	84	74	59	43	39
Property Damage						
$ 5,000	$36	$31	$28	$24	$20	$18
10,000	38	34	30	26	22	20
25,000	41	37	33	29	25	23
50,000	45	40	37	33	28	27

The next three tables are for the other three factors that insurance companies consider when setting the base premium:

TABLE 4
AGE FACTOR TABLE

Male		Female	
Age	*Factor*	*Age*	*Factor*
Under 21	2.50	Under 21	1.65
21–24	1.60	21–up	1.00
25–29	1.50		
30–up	1.00		

TABLE 5
USE OF CAR FACTOR TABLE

Pleasure	Business	Work Less than 20 Miles	Work 20 Miles or More
1.00	1.75	1.15	1.55

TABLE 6
DRIVING RECORD FACTOR TABLE

Number of Accidents	0	1	2	3	4
Factor	0.00	0.25	0.65	1.40	1.75

To find the car insurance premium, Tables 4, 5, and 6 are consulted to get the total factor. Then that factor is multiplied by each of the two premiums (bodily injury and property damage), as in the following examples.

EXAMPLE 1

Harry's car is insured by Ajax Car Insurance Co. He is 22. He wants a 25/50 bodily injury policy with $25,000 property damage coverage. He lives in territory 03 and had one accident last year. He uses the car for pleasure. Calculate his insurance premium.

SOLUTION

First consult Tables 4, 5, and 6 to find the factors for age, use of car, and driving record:

Age factor:	1.6
Use of car:	1.0
Driving record:	0.25
Total Factor:	2.85

Now find the two premiums:

Bodily injury (25/50, territory 03): $66

Property damage ($25,000, territory 03): $33

Now multiply the total factor by each of the premiums and add them to get the total premium on the car:

$$66 \times 2.85 = \$188.10$$
$$33 \times 2.85 = \$94.05$$
$$\$188.10 + \$94.05 = \$282.15$$

The premium on Harry's car will be $282.15.

EXAMPLE 2 Lillian had her car insured by Acme Insurance, Inc. She is 25. She wants a 100/300 bodily injury policy with $50,000 property damage coverage. She lives in territory 05 and has had no accidents in the last three years. She uses the car to get to work, and work is less than 20 miles away. Calculate her insurance premium.

SOLUTION

First consult Tables 4, 5, and 6 to find the factors for age, use of car, and driving record:

Age factor: 1.0

Use of car: 1.15

Driving record: 0.0

Total Factor: 2.15

Now find the two premiums:

Bodily injury (100/300, territory 05): $43

Property damage ($50,000, territory 05): $28

Now multiply the total factor by each of the premiums and add them to get the total premium on the car:

$$43 \times 2.15 = \$92.45$$
$$28 \times 2.15 = \$60.20$$
$$\$92.45 + \$60.20 = \$152.65$$

Lillian will pay $152.65 to insure her car.

EXERCISES

Exercise A   Using Tables 3 to 6, find the premium for each of the following bodily injury insurance coverages:

	Coverage	Territory	Sex	Age	Use	Number of Accidents	Premium	
1.	10/20	06	F	19	Pleasure	1	_____	____
2.	25/50	01	M	26	Business	0	_____	____
3.	50/100	05	F	30	Under 20 miles	0	_____	____

Find the premium for each of the following property damage insurance coverages:

	Coverage	Territory	Sex	Age	Use	Number of Accidents	Premium	
4.	$ 5,000	06	F	20	Pleasure	0	_____	____
5.	25,000	04	M	26	Over 20 miles	0	_____	____

Word Problems

Exercise B   Solve the following problems:

6. Arthur Wilson carries $100,000/$300,000 bodily injury coverage and $25,000 property damage coverage. He is 27 years old, lives in territory 02, and drives to work a distance of 17 miles. He had one accident 2 years ago. What are his yearly premiums for both coverages?

7. During a rainstorm Luis skidded off the road and hit a tree. Repairs on his car amounted to $348.75. If he carried $50-deductible collision insurance coverage, how much did the insurance company pay for the damages?

8. Evelyn is 20 years old, lives in territory 03, and drives to work a distance of 23 miles. She had one accident 2 years ago. Her coverage is $100,000/$300,000 for bodily injury and $50,000 for property damage.

   (a) Find the yearly premium for both coverages.
   (b) When she renews her policy next year, she will be over 21. What will her premium be then?

Unit 5: Life Insurance

If a family member dies, **life insurance** will compensate the family financially for the **face value** of the insurance policy (the amount of money to be paid). This becomes especially important if the deceased member is the major breadwinner of the family. However, life insurance is often taken out on each member of the family. As with other insurance policies, the amount paid every month is called the **premium**, and the person(s) being insured is (are) called the **policyholder(s)**. The person who receives the insurance payment after the policyholder has died is the **beneficiary**. There are five major kinds of life insurance:

1. **Straight or whole life.** This actually offers two kinds of benefits. In the event of the death of the policyholder, the family will receive the face value of the policy. In addition, the policy builds up equity over time. That money can be drawn on once the policyholder has retired. It is similar to a retirement savings account in this way.

2. **Limited payment.** This is similar to a whole-life policy, except that the payments are made over a specified time span (20–30 years is typical). During the payment period, the policyholder is covered for the face value of the policy. After the specified time is up, the policyholder stops making payments but is covered for the rest of his/her life. The beneficiary will receive the face value of the policy in the event of the policyholder's death.

3. **Term life.** This is much less expensive than the straight or whole-life policy but is a simple life insurance policy. As long as the premiums are paid, the beneficiary will receive the face value on the policy. Once the premiums are stopped, the policy is canceled. No equity is built up over time.

4. **Endowment.** This is a lot like a limited-payment policy except that at the end of the payment period, the policyholder will receive the face value of the policy or an annuity. An endowment policy is one of the more-expensive policies available.

5. **Universal life.** This is similar to a straight or whole-life policy except that the premiums are variable. They fluctuate with the prime lending rate, stock and bond markets, and other factors. It is cheaper than whole-life policies but is somewhat riskier.

TABLE 7
ANNUAL PREMIUMS PER $1,000 OF LIFE INSURANCE

Age at Issue	Term		Straight Life	Limited-Payment Life		Endowment	
	10 Years	*15 Years*		*20 Years*	*30 Years*	*20 Years*	*30 Years*
20–24	$ 8.44	$ 9.21	$19.72	$34.63	$26.82	$57.42	$46.72
25–29	9.33	10.05	22.68	38.24	31.24	58.92	48.32
30–34	10.42	11.78	25.35	42.38	35.62	61.42	51.60
35–39	12.73	14.61	29.30	46.71	39.80	63.19	53.90
40–44	15.82	18.19	35.84	50.28	44.21	65.38	55.34
45–49	20.47	23.68	39.95	55.65	49.62	68.17	57.43
50–54	27.38	32.28	49.10	62.28	56.34	71.24	_____
55–59	38.64	_____	60.18	69.08	63.19	73.82	_____
60+	_____	_____	75.72	81.54	73.34	74.15	_____

Oscar has a limited-payment life insurance policy from New Gretna Life Insurance Co., Inc., with a face value of $70,000. He was 25 when he took out the policy. He chose the 30-year policy. Calculate his annual premium.

SOLUTION

First consult Table 7. Look up age 25, limited-payment policy at 30-years. The premium is $31.24.

Now find the number of thousands in the face value:

$$70,000/1,000 = 70$$

Now multiply the premium by the number of thousands:

$$70 \times 31.24 = 2,186.8$$

The annual premium Oscar will pay for a 30-year, limited-payment policy is $2,186.80

EXERCISES

Exercise A   Using Table 7, find the annual premium for each of the following policies:

	Policy	Age at Issue	Face Value	Premium per $1,000		Number of 1,000's in Face Value	Annual Premium	
1.	15-year term	22	$22,500	___	___	___	___	___
2.	Straight life	28	24,500	___	___	___	___	___
3.	10-year term	26	19,500	___	___	___	___	___
4.	Straight life	33	40,000	___	___	___	___	___
5.	10-year term	31	29,000	___	___	___	___	___

Word Problems

Exercise B   Solve the following problems:

6. How much will the annual premium be on a 15-year term policy with a face value of $28,500, issued at age 26?

7. At age 40, Gus Hartman purchased a 30-year endowment policy with a face value of $25,500. How much less would his annual premiums have been if he had purchased the policy at age 30?

8. Ernesto Hernandez purchased a $32,500 20-year endowment policy at age 34. How much will he have paid in premiums for the 20-year period?

Unit 6: Review of Chapter 7

Using Table 1, find the monthly and yearly payments for each of the following 30-year mortgages:

	Mortgage	Rate	Monthly Payment		Yearly Payment	
1.	$46,750	$10\frac{1}{2}\%$	___	___	___	___
2.	53,250	$11\frac{1}{2}$	___	___	___	___

Find the real estate tax on each of the following properties:

	Assessed Valuation	Tax Rate	Amount of Tax	
3.	$24,700	$8.63 per $100	_____	___
4.	23,500	$11.25 per $1,000	_____	___

Find the premiums for 1 year in each of the following policies:

	Face Value of Policy	Annual Rate per $100	Annual Premium	
5.	$27,850	$0.47	_____	___
6.	39,250	0.48	_____	___

Find the total premium on the following home insurance policy for the term indicated. (Use Table 2 on page 194.)

	Face Value of Policy	Annual Rate per $100	Term of Policy	Total Premium	
7.	$37,750	$0.46	5 years	_____	___

Find the number of unexpired days and the amount of refund on the following canceled yearly policies:

	Annual Premium	Policy in Force	Unexpired Days	Amount of Refund	
8.	$235	113 days	_____	_____	___
9.	253	195 days	_____	_____	___

Find the annual depreciation for the following car:

	Original Cost	Trade-In Value		Annual Depreciation	
		At the End of	Amount		
10.	$7,990	6 years	$1,950	_____	___

Find the annual depreciation and the rate of depreciation to the nearest percent in each of the following:

	Original Cost	Trade-In Value		Annual Depreciation	Rate of Depreciation
		At the End of	Amount		
11.	$9,890	5 years	$2,725	_____	____
12.	9,745	5 years	2,650	_____	____

Find the premium for each of the following car insurance coverages:

	Coverage	Territory	Sex	Age	Use	Number of Accidents	Premium
13.	Bodily injury 100/300	05	M	28	Over 20 miles	3	_____ ____
14.	Property damage $25,000	04	M	32	Under 20 miles	2	_____ ____

Find the annual premiums on each of the following life insurance policies:

	Policy	Age at Issue	Face Value	Premium per $1,000	Number of 1,000's in Face Value	Annual Premium
15.	15-year term	24	$52,750	____ ____	_____	_____ ____
16.	30-year endowment	34	70,000	____ ____	_____	_____ ____

Word Problems

Solve the following problems:

17. Adrienne Ganz bought a condominium for $66,250. She made a down payment of 20% and secured a 30-year mortgage for the balance at $13\frac{1}{2}$% interest.

    (a) How much will her monthly payments be?
    (b) How much less money will she pay for the same mortgage over a period of 25 years instead of 30 years?

18. Howard paid $67,500 for his house. His tax rate is $95.25 per $1,000, and his house is assessed at 50% of the purchase price.

    (a) How much will he pay per year in real estate taxes?
    (b) If he sold his house for $105,500 ten years later, how much will the new buyer pay a year in real estate taxes?

The Mathematics of Retailing

Unit 1: Sales Slips

The sales slip records a sale of goods or services. Although most times a cash register (or computer cash register) will print out a **register receipt**, handwritten **sales slips** are still common in business.

Most sales slips should contain all of these parts:

1. The name, address, and telephone number of the business offering the sale or the service. This is usually printed on the slip.

2. The date of the purchase and the name and address of the purchaser.

3. The quantity of the items purchased, a description of the items, the unit price of each, and the amount.

4. The total amount of the purchase (found by adding up the totals of the individual items).

5. The exchange policy of the store, which clearly spells out the conditions for returning and/or exchanging merchandise.

The slip on the left is a properly filled-out sales slip. On the right is a cash register slip for the same purchases.

<table>
<tr><td colspan="4" align="center">

G & G CLOTHING CO.
372 Elm Street
Baltimore, MD 21204

Aug. 3, 20___

SOLD TO   A. Borden

ADDRESS   475 Macon Street

Baltimore, MD 21204
</td></tr>
</table>

Quantity	Description	Unit Price	Amount
4	Shirts	5 75	23 00
2 pr.	Slacks	18 50	37 00
4 pr.	Socks	1 50	6 00
			66 00
		Sales tax	5 28
		Total	71 28

REFUNDS AND EXCHANGES MADE WITHIN
7 DAYS FROM DATE OF PURCHASE.

G & G CLOTHING CO.

Aug. 03, 20–

	5.75
	5.75
	5.75
	5.75
	18.50
	18.50
	1.50
	1.50
	1.50
	1.50
Subtotal	66.00
Tax	5.28
Total	71.28

EXERCISES

Exercise A   Find the total amount of each of the following sales. (The symbol @ means "each." Thus, "3 shirts @ $5.95" means that each shirt costs $5.95.)

1. 3 White shirts @ $5.95   _____
   2 pr. Slacks @ $17.50   _____
   1 Sport jacket @ $39.75   _____

                             _____
     8% Sales tax →   _____
         Total   _____

2. 3 pr. Jeans @ $12.50   _____
    2 pr. Shoes @ $14.95   _____
    5 pr. Socks @ $1.95   _____
    1 pr. Sneakers @ $11.80   _____

                                   _____
    7% Sales tax →   _____
    Total   _____

3. 4 Window shades @ $6.25   _____
    3 Curtains @ $11.90   _____
    3 Curtain rods @ $3.75   _____

                                   _____
    3% Sales tax →   _____
    Total   _____

Exercise B   Using the forms provided, write a sales slip for each of the following purchases:

4. 2 bathing suits @ $27.95; 1 beach robe @ $15.50; 2 beach towels @ $3.75; 2 bathing caps @ $4.50. The sales tax is 5%.

5. 5 sets of underwear @ $3.50; 8 pairs of socks @ $1.75; 2 pairs of slippers @ $5.95. The sales tax is 3%.

4.

20

SOLD TO _____

ADDRESS _____

CLERK	DEPT	AMT REC'D

QUAN.	DESCRIPTION	AMOUNT
	Subtotal	
	8% Sales tax	
	Total	

POSITIVELY NO EXCHANGES MADE UNLESS
THIS SLIP IS PRESENTED WITHIN 3 DAYS

5.

20

SOLD TO _____

ADDRESS _____

CLERK	DEPT	AMT REC'D

QUAN.	DESCRIPTION	AMOUNT
	Subtotal	
	8% Sales tax	
	Total	

POSITIVELY NO EXCHANGES MADE UNLESS
THIS SLIP IS PRESENTED WITHIN 3 DAYS

Unit 2: Unit Pricing

A shopper can be confused by the great number of sizes of goods in stores these days. It can be difficult to determine whether a 12-ounce can of tomato juice for 59 cents is a better or a worse buy than an 18-ounce can of the same tomato juice for $1.29. For this reason, stores have adopted **unit pricing**. With unit pricing, consumers can compare prices and determine which is the best price for the size of the goods sold. Unit pricing is required by law in many areas. With unit pricing, two prices are displayed. One is the price of the goods and the size. The other is the unit price for a standard size (a pound or 100 count, for instance).

This system works pretty well. However, sometimes the unit pricing can be confusing. The problem is that some stores will give two or more "standard units" for the unit pricing, and the result can be that the unit pricing can be as confusing as the regular price. For instance, it has been observed that in the case of juice, a unit price for a pound will be shown in front of one size and a unit price for a quart will be shown in front of another size. It is vital, therefore, to double-check the standard units and then standardize the units to determine which is the better buy. Generally speaking, the larger the size, the more you get for your money for a certain commodity. This does not *always* work that way, however. Checking the unit pricing can help you determine which is the better buy.

To determine the unit price, decide what unit you want to use. This is an important decision as it drives your solution path. Usually, you will use pounds and ounces for dry goods. You will use gallons, quarts, pints, and ounces for liquid goods. (Be careful *not* to confuse dry ounces with liquid ounces!) You will use inches, feet, and yards for linear measure (for cloth or lumber sold by the foot, for instance). You should use multiple units for pills or other commodities that are sold in small units (100 pills for $2.99, for instance).

Next, you should decide on your solution path. Most of the time, although not always, this will reduce the commodity to a unit price. For instance, let's consider the 12-ounce can of tomato juice for 59 cents and the 18-ounce can for 1.29 described earlier. Divide 59 by 12:

$$59/12 = 4.91$$

This size can costs 4.9 cents per ounce. Now divide 129 (cents) by 18:

$$129/18 = 7.1$$

This size can costs 7.1 cents per ounce. You can now see that the smaller can is a better buy than the larger can.

At other times, it might be best to calculate the price for a larger unit (gallons, for instance). With small commodities, such as pens, it might be easiest calculate the cost for a dozen pens. With very small commodities (pills, for instance), it is often easiest to find the price per 100 pills because most pills come in a 100-count bottle.

In most cases, dividing the price by the unit, or multiplying the price by units to get a larger unit, is the easiest way to calculate unit pricing. Example 2 below shows this.

In some cases, it might be easiest to set up a proportion. For instance, let's take potatoes. They sell for $2.99 for 5 pounds or $1.59 for 2 pounds. Which is the better buy? Set up two proportions:

$$\frac{2.99}{5} = \frac{x}{1} \quad \text{and} \quad \frac{1.59}{2} = \frac{x}{1}$$

Solving these for x gives:

$$5x = 2.99(1) \Rightarrow x = 0.598 \quad 2x = 1.59(1) \Rightarrow x = 0.795$$

We see that buying 5 pounds of potatoes is the better buy. Let's look at some other examples. Use the following table to help your calculations:

Standard Units of Measure

Weight	Units
1 pound = 16 ounces	1 dozen = 12 items
1 ton = 2,000 pounds	1 gross = 144 items (12 dozen)
Distance	**Liquids**
1 foot = 12 inches	1 pint = 16 ounces
1 yard = 3 feet (36 inches)	1 quart = 2 pints (32 ounces)
	1 gallon = 4 quarts

 EXAMPLE 1   Which is the better buy, 80 pills for $2.98 or 250 pills for $5.35?

SOLUTION

Set up proportions to calculate both for 100 pills:

$$\frac{2.98}{80} = \frac{x}{100} \quad \text{and} \quad \frac{5.35}{250} = \frac{x}{100}$$

Solving for x gives:

$$80x = (2.98)(100) \Rightarrow x = 298/80 \Rightarrow x = 3.725$$

The 80-pill bottle is $3.73 for 100.

$$250x = (5.35)(100) \Rightarrow x = 535/250 \Rightarrow x = 2.14$$

The 250-pill bottle is $2.14 for 100. We see that the 250-pill bottle is the better buy.

EXAMPLE 2
At the Shaheen Paint Store, Donovan's Paint costs for $5.99 per quart or $24.99 for a gallon. Which is the better buy?

SOLUTION

To solve this, we will find the cost of each per gallon. The second one is per gallon already:

$$\frac{5.99}{1} = \frac{x}{4} \quad (1)x = (5.99)(4) \Rightarrow x = 23.96$$

The quart price gives us $23.96 for a gallon. However, the gallon costs $24.99. So the paint is cheaper buying it by the quart.

 EXAMPLE 3
At Scrantom's Stationery Store, you can buy pens 3 for a dollar or 4 for $1.25. Which is the better buy?

SOLUTION

Here is an example where using a larger unit, like a dozen, might be easier. Set up the proportions:

$$\frac{1.00}{3} = \frac{x}{12} \quad \text{and} \quad \frac{1.25}{4} = \frac{x}{12}$$

Solving for x gives:

$$3x = (1.00)(12) \Rightarrow x = 12/3 \Rightarrow x = 4$$

These pens cost $4.00 for 12 (a dozen).

$$4x = (1.25)(12) \Rightarrow x = 15/4 \Rightarrow x = 3.75$$

These pens cost $3.75 for 12 (a dozen). The pens selling for 4 for $1.25 is the better deal.

 EXAMPLE 4
Bananas are selling 3 pounds for a dollar. What is the unit price?

SOLUTION
We simply divide a dollar (1.00) by 3 (pounds).

$$1.00/3 = 0.333$$

Bananas are selling for 33 cents per pound.

Rope is selling at $4.75 for a yard. How much does it cost per foot?

SOLUTION

There are 3 feet to a yard:

$$\frac{4.75}{3} = \frac{x}{1} \quad (3)x = (4.75)(1) \Rightarrow x - 4.75/3 \Rightarrow x = 1.583$$

The rope sells for $1.58 per foot.

 Salami is selling for $4.99 per pound. How much is it selling per ounce?

SOLUTION

A proportion will work here (1 pound equals 16 ounces):

$$\frac{4.99}{16} = \frac{x}{1} \quad (16)x = (4.99)(1) \Rightarrow x = 4.99/16 \Rightarrow x = 0.3118$$

The salami sells for 31 cents per ounce.

 A customer wants to buy 6 ounces of potato salad that sells for $3.75 a pound. How much will the customer pay?

SOLUTION

A proportion is the easiest method to use here. Remember that 16 ounces equal 1 pound:

$$\frac{3.75}{16} = \frac{x}{6} \quad (16)x = (3.75)(6) \Rightarrow x = 22.50/16 \Rightarrow x = 1.406$$

The customer will pay $1.41 for the potato salad.

EXERCISES

Exercise A   From the information given, find the unit price of each of the following items:

	Item	Fractional Weight or Measure	Cost of Fractional Unit	Unit	Price per Unit	
1.	Candy	8-oz. bag	89¢	Pound	____	____
2.	Ribbon	2 ft.	62¢	Yard	____	____
3.	Pencils	36 doz.	$2.75	Gross	____	____
4.	Wine	3 qt.	6.75	Gallon	____	____
5.	Socks	8 pr.	10.40	Dozen	____	____

For each of the following items, from the information given, find the fractional part and its cost.

	Item	Price per Unit	Quantity Purchased	Fractional Part	Cost of Fractional Part	
6.	Coffee	$5.35 per pound	12 oz.	____	____	____
7.	Gingham	3.49 per yard	28 in.	____	____	____
8.	Ribbon	2.15 per yard	2 ft.	____	____	____
9.	Velvet	6.25 per yard	48 in.	____	____	____
10.	Steak	2.85 per pound	25 oz.	____	____	____

Word Problems

Exercise B   Solve the following problems:

11. Mr. Welk bought a 6-oz. bag of candy for $1.45.

   (a) What fraction of a pound did he buy?
   (b) What is the price per pound?

12. Mrs. Acosta bought $\frac{2}{3}$ yard of satin for $2.49. What is the price per yard?

13. Mr. Goldberg bought 48 dozen wood screws for $12.75. What is the price per gross?

14. How much would 14 oz. of cheese cost if the price was $2.69 per pound?

15. Geraldine bought 18 hair rollers priced at $4.89 per dozen. How much did she pay?

Unit 3: Sale Items

When items are offered at the normal price, it is often referred to as the **list price** (or the **marked price** or the **selling price**). When an item goes on sale, the amount of reduction in price is sometimes referred to as the **discount** or **markdown**. The new price, the **sale price**, is sometimes put next to the list price for comparison. There are lots of reasons for sales. Slow-moving items will be sold at a discount. At the end of a season, a retailer will want to clear out items for the next season. In August, for instance, swimsuits and other summer apparel will go on sale to make way for the fall clothing. Floor samples can be deeply discounted because they were handled a lot. The two common types of reduction are percent and dollar amount. Dollar reduction is where the sale price is a certain number of dollars (or cents) off the list price not necessarily a percent. Percent reduction is where the sale price is a percent of the list price. We will look at percent reductions in this unit.

CALCULATING THE SALE PRICE

Two calculations are important when calculating the sale price: the **discount amount**, and the **sale price**. When an item is reduced in price through a discount percent, the percent is multiplied by the list price to get the discount amount. Then the discount amount is subtracted from the list price to get the sale price, as in these examples.

 EXAMPLE 1   At Frank's Big 125 Furniture Warehouse, sofas sell for $499.00. During his Spring Sale, however, Frank is discounting the price 30%. Find the discount amount and the sale price.

SOLUTION

First calculate the discount amount by multiplying the discount percent by the list price:

$$499 \times 0.30 = 149.7$$

The discount amount is $149.70.

Then subtract the discount amount from the list price:

$$499 - 149.70 = 349.30$$

The sale price is $349.30.

Broadway Camera sells Mortar Digital Cameras for $739.95. They had their Annual Fire Sale and discounted the price 15%. Find the discount amount and the sale price.

SOLUTION

First calculate the discount amount by multiplying the discount percent by the list price:

$$739.95 \times 0.15 = 110.99$$

The discount amount is $110.99.

Then subtract the discount amount from the list price:

$$739.95 - 110.99 = 628.96$$

The sale price is $628.96.

Star Markets sells bottom round roast for $6.99 a pound. During last week's sale, the price was discounted 40%. Find the discount amount and the sale price.

SOLUTION

First calculate the discount amount by multiplying the discount percent by the list price:

$$6.99 \times 0.40 = 2.796$$

The discount amount is $2.80.

Then subtract the discount amount from the list price:

$$6.99 - 2.80 = 4.19$$

The sale price is $4.19 per pound.

Alternatively, you could calculate the sale price directly. If the discount amount is 40%, then the sale price is $100 - 40 = 60\%$ of the price. So just multiply 0.60 by the list price:

$$6.99 \times 0.60 = 4.19$$

The sale price, again, is $4.19 per pound.

Sibley's Department Store sells DVD players for $199.90. During their Holiday Fiesta Days, the price was discounted 25%. Find the discount amount and the sale price.

SOLUTION

First calculate the discount amount by multiplying the discount percent by the list price:

$$199.90 \times 0.25 = 49.975$$

The discount amount is $49.98.

Next subtract the discount amount from the list price:

$$199.90 - 49.98 = 149.92$$

The sale price is $149.92.

As with the previous example, you could calculate the sale price directly. If the discount amount is 25%, then the sale price is $100 - 25 = 75\%$ of the price. So just multiply 0.75 by the list price:

$$199.90 \times 0.75 = 149.92$$

The sale price, again, is $149.92.

EXERCISES

Exercise A   Complete the following tables:

	Item	Original Price		Discount Percent	Discount Amount		Sale Price	
1.	Coat	$125	90	35%	___	___	___	___
2.	TV	169	50	25%	___	___	___	___
3.	Desk	87	50	$\frac{2}{3}$ off	___	___	___	___

	Item	Original Price		Discount Percent	Remaining Percent	Sale Price	
4.	Chair	$115	75	30%	___	___	___
5.	Typewriter	147	50	$\frac{2}{3}$ off	___	___	___

Word Problems

Exercise B   Solve the following problems:

6. A refrigerator selling for $265.75 is marked down 25%. What is the sale price?

7. A kitchen set selling for $342 is marked down 25%. What is the selling price of the set?

8. A lamp selling for $72 was marked down to sell for $54. Find the discount percent.

Unit 4: Review of Chapter 8

Find the total amounts of each of the following sales:
Sample Solution:

4 Shirts @ $9.95	$39.80
2 Ties @ $6.75	13.50
1 Sweater @ $15.90	15.90
	69.20

$69 \boxdot 20 \; \boxtimes \; 0 \boxdot 095 \; \boxminus \; 6.57 \; \underline{4}$
$= \$6.57$

$9\frac{1}{2}$% Sales tax (9.5%)	6.57
Total	$75.77

1.

5 gal. White paint @ $8.85	_____
6 rolls Wallpaper @ $15.95	_____
1 Roller/pan set @ $5.35	_____
2 Brushes @ $8.25	_____
	_____
$7\frac{1}{2}$% Sales tax	_____
Total	_____

From the given information, find the unit price of each of the following items:

	Item	Fractional Weight or Measure	Cost of Fractional Unit		Unit	Price per Unit	
2.	Steak	13.4 oz.	$2	68	Pound	____	____
3.	Ribbon	28 in.	3	60	Yard	____	____

Find the fractional part and the cost of the fractional part for each of the following items:

	Item	Price per Unit		Quantity Purchased	Fractional Part	Cost of Fractional Part	
4.	Veal	$3	85/lb.	$14\frac{1}{2}$ oz.	____	____	____
5.	Wine	6	95/gal.	7 qt.	____	____	____

Find the discount amount and the sale price of the following item:

Item	Original Price		Discount Percent	Discount Amount		Sale Price	
6. A	$178	95	$\frac{1}{3}$ off	_____	____	_____	____

Find the remaining percent and the sale price of each of the following items. (Subtract the discount percent from 100%.)

Item	Original Price		Discount Percent	Remaining Percent	Sale Price	
7. A	$342	85	45%	_____	_____	____
8. B	169	95	$12\frac{1}{2}$	_____	_____	____

Word Problems

Solve the following problems:

9. If ham sells at $4.65 per pound, how much will 6 ounces cost?

10. A piano originally selling for $1,245.75 is on sale at 40% off. What is the sale price of the piano?

The Mathematics of Purchasing

Unit 1: Purchasing Goods

A plastic manufacturer needs oil to make plastic snow shovels. A music store needs to stock guitars, amplifiers, drum sets, and so on. A school needs books and pens to start the year. All types of businesses need to purchase goods to keep their business going.

THE PURCHASE ORDER

When a business or manufacturer (**buyer**) needs to purchase goods, it uses a **purchase order**. The purchase order is a formal, commercial document in which a buyer requests goods (or services) from a **seller** (the business supplying the goods or services). When agreed to, the purchase order becomes a contract between the buyer and the seller.

- The purchase order clearly states the type and amount of goods requested, the price for each item, and the total price for all goods/services.
- It is dated, and it specifies how the goods are to be shipped.
- It specifies the last acceptable date for shipping the goods.
- It specifies how and when the **invoice** is to be paid. (The invoice is explained in the next section.)

These terms are called the **terms of purchase**. All purchase orders have a number on them so they can be referred to more easily. Typically, the name and address of the buyer and the seller will be clearly printed, or typed, as shown.

ACME MANUFACTURING COMPANY
1775 Patriot Way
Burlington, VT 05401

To:

Ajax Box Corporation
243 E Willowdale Ave
Santa Monica, Vt 05040

PURCHASE ORDER

No.   127356
Date   March 3, 20-
Ship via   Truck
Terms   2/10 N30

Quantity	Description	Unit Price	Amount
5000	#170 Corrugated Cartons	0.0375	$187.50
	Discount	10%	- 18.75
	Net		$168.75
	Not for resale		
	Tax exempt No. 12374-36		

In the above purchase order, the terms of purchase are stated as 2/10 N30. This means that the purchaser may pay the bill in one of two ways. Acme manufacturing company can deduct a cash discount of 2% if the total amount is paid in 10 days from the date of the invoice or can pay the full amount within 30 days.

THE INVOICE

Once the goods have been shipped (or the services rendered), the seller sends (or hands) the buyer an **invoice**, This is essentially a bill for the goods or services rendered. The invoice will be dated, numbered, specify what goods or services were rendered, and clearly state the unit and total prices for each item. Sometimes if the seller does not have all the items the buyer wants, the seller will ship a partial order. A partial shipment will be reflected on the invoice. In addition, an invoice contains the terms of purchase, as explained above. The buyer can use the invoice as a check to determine how much of the merchandise was shipped (whether the order was filled completely or only partially). As with the purchase order, the name and address of the buyer and the seller will be clearly printed, or typed, as shown.

AJAX BOX CORPORATION 243 East Willowdale Avenue Santa Monica, VT 05040			INVOICE NO. **24749** DATE March 25, 20 –			

S O L D T O: Acme Manufacturing Corporation
1775 Patriot Way
Burlington, Vt 05401

S H I P T O:

CUST. ORDER NO. 127356	ORDER DATE Mar. 3, 20--	SALESMAN Mail	F.O.B.	TERMS 2/10 n 30		SHIP VIA Truck
NO. OF CASES	PACK	DESCRIPTION		CODE	PRICE	AMOUNT
10	500	#170 Corrugated cartons		17356-X	0.0375	$187 50
		10% Trade discount				– 18 75
						$168 75
		Transportation				+ 14 63
					TOTAL AMOUNT	$183 38

INVOICE

Concerning the terms of service, the **discount date** is the last date on which a buyer can pay the bill to receive the early-payment discount. The **due date** is the last day on which the buyer can pay the bill. After that, the buyer is in default.

 EXAMPLE 1
Find the discount date and the due date of an invoice that is dated March 15 and that has the terms 2/10 n 45.

SOLUTION

The discount date is 10 days past March 15, or March 25. That is the last date on which the buyer can pay the bill and get a 2% cash discount.

The due date is 45 days past March 15. Be careful. March has 31 days. So 30 days past March is April 14. The terms are 45 days, so the due date is 15 days past April 14, or April 29.

Hint: It might be a good idea to consult a calendar when you are working on due dates.

EXAMPLE 2 An invoice dated December 20 has a total bill of $850.50 and has the terms 2/10 n 30. Find:

    a) The discount date
    b) The due date
    c) The cash discount amount if the bill is paid within the discount date
    d) The net amount paid if the cash discount is taken

SOLUTION

    a) The discount date is 10 days past December 20, or December 30. That is the last date on which the buyer can pay the bill and get a 2% cash discount.
    b) The due date is 30 days past December 20. December has 31 days. So 30 days past December 20 is January 19.
    c) The cash discount is 2% of 850.50:

$$850.50 \times 0.02 = \$17.01$$

    d) The net amount paid if the cash discount is taken is:

$$850.50 - 17.01 = \$833.49$$

EXERCISES

Exercise A   Find the discount date and the due date in each of the following:

	Invoice Date	Terms	Discount Date	Due Date
1.	April 5	2/10 n 30	_____	_____
2.	April 25	2/10 n 45	_____	_____
3.	February 11	2/10 n 60	_____	_____
4.	August 23	2/15 n 60	_____	_____
5.	January 13	2/10 n 30	_____	_____

Find the discount date, the due date, and the net amount *or* the full amount as required in each of the following invoices:

	Invoice Date	Terms	Amount of Invoice	Discount Date	Due Date	Invoice Paid	Net Amount		Full Amount	
6.	9/10	2/10 n 45	$ 467 50	_____	_____	9/18	___	___	___	___
7.	7/15	3/15 n 60	392 70	_____	_____	7/30	___	___	___	___
8.	10/8	2/10 n 45	963 85	_____	_____	11/25	___	___	___	___
9.	2/8	2/10 n 90	1,242 50	_____	_____	2/15	___	___	___	___
10.	6/16	3/10 n 60	372 45	_____	_____	6/25	___	___	___	___

Word Problems

Exercise B   Solve the following problems:

11. An invoice in the amount of $827.50 is dated 3/9. The terms are 3/15 n 60.

    (a) What is the last day for the discount?
    (b) When does the invoice become due?
    (c) What is the net amount of the invoice, if paid within the discount date?

12. On 7/21 merchandise was shipped for $639.75, terms 3/10 n 45.

    (a) Up to what date may the discount be taken?
    (b) What is the last date for payment to become due?
    (c) What is the net amount paid if the discount is taken?

13. The Lace Novelty Company received an invoice dated 5/23 for $863.75. The terms are 3/10 n 45.

    (a) Up to what date may the discount be taken?
    (b) When is the full amount due?
    (c) How much will the discount be?

Unit 2: Calculating Trade Discounts

Some manufacturers and wholesalers sell the things they make to retailers who, in turn, sell them to the public. Often, manufacturers will suggest the sales price for the goods. Earlier we called this price the **list price**, but it is also called the **suggested retail price**. The manufacturers sell it to retailers at that price less the **trade discount**. The trade discount is often 50% off the list price. Consumers pay the list price, and retailers pay the manufacturer the **net price** or the **invoice price**. (The net price is the list price minus the trade discount.)

Most manufacturers extend this trade discount. It is usually calculated as a percentage of the list price.

To calculate the trade discount, the percent of the discount is multiplied by the list price to get the amount of the discount. Then this discount amount is subtracted from the list price to get the invoice price, or the list price, as in these examples.

A washer-dryer combination machine lists for $2,550.00 less a trade discount of 45%. Find the amount of the discount and the invoice price of the washer-dryer to the retailer.

SOLUTION

First calculate the discount amount by multiplying the discount percent by the list price:

$$2,550 \times 0.45 = 1,147.50$$

The discount amount is $1,147.50.

Now subtract the discount amount from the list price:

$$2,550 - 1,147.50 = 1,402.50$$

The invoice price is $1,402.50.

If this seems familiar to you, it should. We did this in Chapter 8, Unit 3, "Calculating the Sale Price."

A solid-oak rolltop desk lists for $1,995.00 less a trade discount of 40%. Find the amount of the discount and the invoice price of the rolltop desk to the retailer.

SOLUTION

First calculate the discount amount by multiplying the discount percent by the list price:

$$1,995 \times 0.40 = 798$$

The discount amount is $798.

Now subtract the discount amount from the list price:

$$1,995 - 798 = 1,197.00.$$

The invoice price is $1,197.00.

Alternatively, as with the sale price, you could calculate the invoice price directly. If the discount amount is 40%, then the sale price is $100 - 40 = 60\%$ of the price. So just multiply 0.60 by the list price:

$$1,995 \times 0.60 = 1,197$$

The sale price, again, is $1,197.00.

CALCULATING A SERIES OF DISCOUNTS

A manufacturer may extend several trade discounts to retailers to induce them to buy goods in larger quantities than they normally would. These discounts are called **quantity discounts**. For example, the manufacturer may offer an extra 5% off to a retailer who buys 12 bedroom sets and an extra 10% off if that retailer purchases 24 bedroom sets. A retailer who thus buys at a big discount is able to sell it for a lower price than his/her competitors and therefore has an advantage in the market.

When two (or more) discounts are offered, the invoice price is found by deducting each discount separately. The discounts are not simply added up. Each subsequent discount is figured on the remaining balance after the previous discount has been deducted.

Pear Computers lists their signature computer, the Bartlett (Bart), for $749.50. Bottoms Computer Co. has purchased 30 Barts at a trade discount of 45%. By buying 20, they received a quantity discount of 10%. Since they purchased an additional 10, to make 30, they were extended a second quantity discount of 5%. What is the amount of each discount and the final invoice price to Bottoms Computer Co.?

SOLUTION

First calculate the discount amount by multiplying the trade discount by the list price:

$$749.50 \times 0.45 = 337.275$$

The discount amount is $337.28.

Now subtract the discount amount from the list price:

$$749.50 - 337.28 = 412.22$$

The first price is $412.22.

Now calculate the first quantity discount (10% on $412.22).

$$412.22 \times 0.10 = 41.22 \Rightarrow 412.22 - 41.22 = 371$$

The second price is $371.00.

Now calculate the third discount (5% on 371).

$$371 \times 0.05 = 18.55 \Rightarrow 371 - 18.55 = 352.45$$

After the trade discount and the two quantity discounts, the Bart computers will cost Bottoms Computer Co. $352.45 per computer.

First alternate solution: As with the sale price, you could calculate the invoice price directly.

If the trade discount amount is 45%, then the first price is $100 - 45 = 55\%$ of the price. Just multiply 0.55 by the list price:

$$749.5 \times 0.55 = 412.225$$

The first price is $412.22.

The first quantity discount is 10%, so the second price is $100 - 10 = 90\%$ of the first price:

$$412.22 \times 0.90 = 370.998$$

The second price is $371.00.

The second quantity discount is 5%, so the final price is $100 - 5 = 95\%$ of the second price:

$$371 \times 0.95 = 352.45$$

The final price is $352.45.

Second alternate solution: You could calculate all the discounts at once. To do this, use all the discount percents that we used in the first alternate solution and multiply them together:

$$0.55 \times 0.90 \times 0.95 = 0.47025$$

This percent represents *all* the discounts. Multiply this percent by the list price:

$$749.50 \times 0.47025 = 352.45$$

The final price is $352.45.

EXERCISES

Exercise A   Find the amount of discount and the invoice price for each of the following:

	List Price		Trade Discount Percent	Amount of Discount		Invoice Price	
1.	$245	50	35%	____	__	____	__
2.	135	60	50	____	__	____	__
3.	316	75	45	____	__	____	__
4.	233	50	55	____	__	____	__
5.	693	80	30	____	__	____	__

Find the amounts of the discounts and the invoice price for each of the following:

	List Price	Trade Discount Percents	First Discount		Second Discount		Third Discount		Invoice Price	
6.	$ 87 | 50	35%, 10%, 5%	——	——	——	——	——	——	——	——
7.	135 | 80	50%, 5%, 3%	——	——	——	——	——	——	——	——
8.	140 | 25	30%, 10%, 5%	——	——	——	——	——	——	——	——
9.	248 | 65	30%, 20%, 5%	——	——	——	——	——	——	——	——
10.	363 | 70	35%, 15%, 10%	——	——	——	——	——	——	——	——

Find the invoice price on each of the following items. (Use the direct method of subtracting from 100%.)

	List Price	Trade Discount Percent	Invoice Price	
11.	$462 | 90	45%	——	——
12.	649 | 25	50	——	——
13.	761 | 35	55	——	——
14.	638 | 25	30	——	——
15.	730 | 50	40	——	——

Find the single discount percent that is equal to the series of discounts in each of the following problems:

	Trade Discount Percents	Single Discount Percent
16.	45%, 15%	——
17.	35%, 20%, 5%	——
18.	25%, 75%, 60%	——
19.	45%, 10%, 5%	——
20.	50%, 5%, 5%	——

Find the single percent that is equal to each series of discounts; then find the invoice price.

	List Price		Trade Discount Percents	Single Equivalent Percent	Invoice Price	
21.	$ 72	60	40%, 10%	____	____	____
22.	241	65	40%, 10%, 5%	____	____	____
23.	281	70	35%, 10%, 5%	____	____	____
24.	$317	45	25%, 20%, 10%	____	____	____
25.	473	90	40%, 5%, 5%	____	____	____

Word Problems

Exercise B   Solve the following problems:

26. A color TV set is listed in a catalog for $395.75 less 40% trade discount.

   (a) What is the amount of discount?
   (b) What is the invoice price of the TV?

27. A retailer purchased six watches listed in the catalog for $87.50 and four watches listed for $125.75. If the trade discount is 45%, how much did he pay for all the watches?

28. An appliance store bought a washing machine listed at $285.75 and four dryers listed at $245.80. If the trade discount is 40%, what is the net amount of the invoice?

29. Wholesaler A offers merchandise with trade discounts of 35%, 5%, and 5%. Wholesaler B offers the exact same merchandise with trade discounts of 30%, 10%, and 5%.

   (a) What is the single equivalent discount for each series of discounts?
   (b) Which wholesaler offers a better buy?

30. An invoice dated 4/10 for $875.60 less 45%, 5%, and 5% is paid on 4/24. If the terms are 3/15 n 30, what should the payment be?

Unit 3: Pricing Goods on the Basis of Selling Price

We've established that retailers buy merchandise from the manufacturer and then sell those goods to consumers at a higher price than what they paid for the merchandise. That's how they stay in business. The price retailers pay is called the **cost**. (In the last unit, remember, it was called the net price or the invoice price.) To cover their costs and make a profit to stay in business, retailers add an amount (sometimes bringing the price up to list price, sometimes not) to their cost. This amount is called the **markup** or the **gross profit**. This markup is the difference between the cost (invoice price) and the selling price.

As mentioned in previous sections, the markup is usually calculated as a percentage of the cost (invoice price). The general practice among retailers is to calculate the markup as a percent as part of their overall business plan. They calculate the markup based on their operating expenses, overhead, and need to make a reasonable profit. This is how they stay in business. It is also, generally, done at the highest levels of the business, by a vice president or general manager, for instance.

The mathematics of pricing retail goods can be expressed in the following fundamental formula:

$$\text{Selling price} = \text{Cost} + \text{Markup}$$

This formula, when transposed, becomes these other two formulas:

$$\text{Markup} = \text{Selling price} - \text{Cost}$$
$$\text{Cost} = \text{Selling price} - \text{Markup}$$

If a retailer pays $150 for an item and sells it for $290, the markup is $140.

FINDING THE AMOUNT OF MARKUP

Finding the amount of markup is identical to the sections on calculating the sale price and calculating the trade discounts, previously covered in this book. The formula is:

$$\text{Selling price} \times \text{Markup percent} = \text{Markup amount}$$

 EXAMPLE 1   Nicolo's Bakery sells their famous twist bread loaves for $3.75. If their markup rate is 35%, what is the markup amount? What is the actual cost of the materials of the bread?

SOLUTION

Multiply the selling price by the markup percent:

$$3.75 \times 0.35 = 1.3125$$

The markup amount is $1.31.

That means that the cost of the materials is:

$$\$3.75 - \$1.31 = \$2.44$$

Alternatively, use the proportion method:

$$\frac{35}{100} = \frac{x}{3.75} \ (100)x = (35)(3.75) \Rightarrow x = 131.25/100 \Rightarrow x = 1.31$$

Again, the cost of the materials is $3.75 − $1.31 = $2.44.

FINDING THE RATE OF MARKUP

In some cases, the markup amount will be given and you will need to calculate the rate (percent) of markup.

Timely Clothes sells a three-piece suit for $399.98. The markup amount for this suit is $159.98. What is the markup rate (percent)?

SOLUTION

The formula is selling price × markup percent = markup amount. By rearranging this formula, we have markup amount/selling price = markup percent.

$$\frac{\$159.98}{\$399.98} = x \Rightarrow x = 0.3999699 \Rightarrow x = 0.40$$

The markup rate (percent) is 40%.

Alternatively, use the proportion method:

$$\frac{\$159.98}{\$399.98} = \frac{x}{100} \Rightarrow (399.98)x = (159.98)(100) \Rightarrow x = 1,5998/399.98$$
$$\Rightarrow x = 39.999 \Rightarrow x = 40$$

Again, we get 40% for the markup rate (percent).

Crazy Harry sells MP3 players at prices that are insane! He sells a high-end MP3 player for $79.90. The markup amount for this device is $27.97. What is the markup rate (percent)?

SOLUTION

The formula is selling price × markup percent = markup amount. By rearranging this formula, we have markup amount/selling price = markup percent.

$$\frac{\$27.97}{\$79.90} = x \Rightarrow x = 0.350062 \Rightarrow x = 0.35$$

The markup rate (percent) is 35%.

Alternatively, use the proportion method:

$$\frac{\$27.97}{\$79.90} = \frac{x}{100} \Rightarrow (79.90)x = (27.97)(100) \Rightarrow x = 2{,}797/79.97$$
$$\Rightarrow x = 35.0062 \Rightarrow x = 35$$

Again, we get 35% for the markup rate (percent).

SELLING PRICE BASED ON MARKUP

Sometimes the retailer needs to set the price based on the markup rate (percent).

Jim Clements is proprietor of The Beat, the best music store in town. He gets a drum set in that cost him $1,550.00 (from the manufacturer). He marks up his merchandise 40%. What price will he set for the drum set?

SOLUTION

First Jim marks up his merchandise 40%, so the cost (to him) is $100 - 40 = 60\%$. The cost represents 60% of the selling price of the drum set. That means that we are asking, $1,550 is 60% of what number? We then set up the proportion:

$$\frac{1{,}550}{x} = \frac{60}{100} \Rightarrow (60)x = (1{,}550)(100) \Rightarrow x = 155{,}000/60 \Rightarrow x = 2{,}583.3333$$

Jim will set the price of the drum set at $2,583.33. Alternatively, we could use:

$$\text{Selling price} \times \text{Cost percent} = \text{Cost}$$
$$(0.60)x - 1{,}550 \rightarrow x = 1{,}550/0.60 \rightarrow x = 2{,}583.333$$

Again, the selling price is $2,583.33.

Down at Handle With Care, a fine glassware and dining room supply store, a shipment of crystal glass vases has come in. The manufacturer charged the store $85.00 (cost) for each of these vases. The store marks up the vases 45%. What price will the store set for the vases?

SOLUTION

First, the store marks up its merchandise 45%. The cost to the store is $100 - 45 = 55\%$. The cost represents 55% of the selling price of the vases. That means that we are asking, $85 is 55% of what number? We then set up the proportion:

$$\frac{85}{x} = \frac{55}{100} \Rightarrow (55)x = (85)(100) \Rightarrow x = 8{,}500/55 \Rightarrow x = 154.5454$$

The store will set the price of the vase at $154.55. Alternatively, we could use:

$$\text{Selling price} \times \text{Cost percent} = \text{Cost}$$
$$(0.55)x = 85 \Rightarrow x = 85/0.55 \Rightarrow x = 154.5454$$

Again, the selling price is $154.55.

Exercise A Find the markup amount and the cost price for each of the following:

	Selling Price		Markup Percent	Markup Amount		Cost Price	
1.	$135	00	35%	____	___	____	___
2.	37	85	45	____	___	____	___
3.	87	50	45	____	___	____	___
4.	120	75	35	____	___	____	___
5.	163	80	45	____	___	____	___

Complete the following table, supplying the cost *or* markup amount *and* the markup percent for each item. (Markup percent is based on selling price.)

	Selling Price		Cost Price		Markup Amount		Markup Percent
6.	$106	25	____	___	$38	50	___
7.	63	90	$ 42	60	____	___	___
8.	138	90	93	95	____	___	___
9.	319	25	143	65	____	___	___
10.	191	45	____	___	85	75	___

Find the equivalent cost percent and the selling price in each of the following. (Markup percent is based on selling price.)

	Cost Price		Markup Percent	Equivalent Cost Percent	Selling Price	
11.	$ 93	65	45%	____	____	___
12.	68	75	40	____	____	___
13.	365	50	55	____	____	___
14.	86	35	30	____	____	___
15.	188	25	25	____	____	___

Word Problems

Exercise B   Solve the following problems:

16. A jeweler pays $95.75 for a ring and sells it for $185.

    (a) What is the markup amount?
    (b) What is the markup percent based on the selling price?

17. A camera store marks up all its merchandise 40% of the selling price. If the cost of a 35 mm camera is $175.50:

    (a) What should the selling price of the camera be?
    (b) What is the markup amount?

18. A shoe store operates on a markup of 45% of the selling price. What is the cost of a pair of shoes that sells for $34.95?

Unit 4: Pricing Goods on the Basis of Cost

Many retailers calculate their markup on a percent of the selling price. Some, however, base their selling price on a percent of the cost (invoice) price. This is similar to the method we looked at in unit 3. The difference is that the cost (invoice) price is used to calculate the markup instead of the selling price.

FINDING THE AMOUNT OF MARKUP

The formula is:

$$\text{Cost} \times \text{Markup percent} = \text{Markup amount}$$

 EXAMPLE 1   Bob's Baby Bottom Store sells a radio to monitor a baby from another room in the house. Bob's cost (invoice price) for each radio is $35.50. His markup is 30% of the cost price. Find the markup amount and the selling price.

SOLUTION

To find the markup amount, solve the equation:

$$\text{Markup amount} = \text{Cost} \times \text{Markup percent}$$
$$x = 35.50 \times 0.30$$
$$x = 10.65$$

The markup amount is $10.65. To find the selling price, solve the equation:

$$\text{Selling price} = \text{Cost price} + \text{Markup amount}$$
$$= \$35.50 + \$10.65$$
$$= \$46.15$$

Alternatively, we could use:

$$Selling\ price = Cost\ price + Markup\ amount$$
$$= \$35.50 + (0.3 \times 35.50)$$
$$= \$35.5 + \$10.65$$
$$= \$46.15$$

FINDING THE RATE OF MARKUP

When you are given the cost (invoice price) and the amount of the markup, you can find the markup rate by rearranging the formula this way:

$$Markup\ percent = Markup\ amount/Cost$$

Walker Brush Co. sells a special brush to photo companies that keeps the emulsion drum clean. Mr. Walker's cost (invoice price) for the brush is $67.50. His markup amount is $21.00. Find the markup rate.

SOLUTION

Solve the equation:

$$Markup\ percent = Markup\ amount/Cost$$
$$x = 21/67.50$$
$$x = 0.3111$$

The markup rate is 0.31, or 31%.

Alternatively, we can ask, $21.00 is what percent of $67.50?

$$\frac{21}{67.5} = \frac{x}{100} \Rightarrow (67.5)x = (21)(100) \Rightarrow x = 2{,}100/67.5$$
$$\Rightarrow x = 31.11 = 31\%$$

Desks Unlimited sells many kinds of desks, chairs, shelves, and other office equipment. The cost (invoice price) for the CEO desk is $2,100.00. The markup amount on this desk is $590.00. Find the markup rate.

SOLUTION

Solve the equation:

$$Markup\ percent = Markup\ amount/Cost$$
$$x = 590/2{,}100$$
$$x = 0.28095$$

The markup rate is 0.28, or 28%.

Alternatively, we can ask, $590 is what percent of $2,100?

$$\frac{590}{2,100} = \frac{x}{100} \Rightarrow (2,100)x = (590)(100) \Rightarrow x = 59,000/2,100$$
$$\Rightarrow x = 28.09 = 28\%$$

SELLING PRICE BASED ON MARKUP

Many retailers will offer a line of merchandise that sells at a specific price. This practice is known as **price lining**.

We use the formula:

$$\text{Cost} + \text{Markup (rate)} = \text{Selling price}$$

Gamrod's sells wallpaper and other home improvement equipment. They sell a roll of wallpaper for $45.50 (selling price). The markup rate on the roll is 55% of the cost (invoice price). Find the cost (invoice price) of the roll of wallpaper.

SOLUTION

The cost price is 100%. The markup rate is 55%, and the selling price is $45.50. Then we have:

$$\text{Cost} + \text{Markup (rate)} = \text{Selling price}$$
$$100\% + 55\% = \$45.50$$
$$155\% = \$45.50$$

The equation to get the cost price is:

$$(1.55)x = \$45.50 \Rightarrow x = 45.5/1.55 \Rightarrow x = 29.354$$

The cost is $29.35.

Alternatively, we can ask, $45.50 is 155% of what number?

$$\frac{45.5}{x} = \frac{155}{100} \Rightarrow (155)x = (45.5)(100) \Rightarrow x = 4,550/155$$
$$\Rightarrow x = 29.354 = \$29.35$$

Marie's Jewelers sells the Silent Service watch for $2,459.95 (selling price). The markup rate on the watch is 60% of the cost (invoice price). Find the cost (invoice price) of the watch.

SOLUTION

The cost price is 100%. The markup rate is 60%, and the selling price is $2,459.95. Then we have:

$$\text{Cost} + \text{Markup (rate)} = \text{Selling price}$$
$$100\% + 60\% = \$2,459.95$$
$$160\% = \$2,459.95$$

The equation to get the cost price is:

$$(1.60)x = \$2{,}459.95 \Rightarrow x = 2{,}459.95/1.60 \Rightarrow x = 1{,}537.468$$

The cost is \$1,537.47.

Alternatively, we can ask, \$2,459.95 is 160% of what number?

$$\frac{2{,}459.95}{x} = \frac{160}{100} \Rightarrow (160)x = (2{,}459.95)(100) \Rightarrow x = 245{,}995/160$$
$$\Rightarrow x = 1{,}537.468 = \$1{,}537.47$$

EXERCISES

Exercise A   Complete the following tables. The markup is based on the cost price.

	Cost Price		Markup Percent	Markup Amount		Selling Price	
1.	$ 48	75	75%	____	___	____	___
2.	93	80	55	____	___	____	___
3.	123	25	65	____	___	____	___
4.	110	90	75	____	___	____	___
5.	165	45	55	____	___	____	___

	Cost Price		Markup Amount		Selling Price		Markup Rate Based on Cost
6.	$137	50	$ 69	38	____	___	____
7.	215	35	____	___	$355	50	____
8.	____	___	51	40	144	85	____
9.	263	50	171	28	____	___	____
10.	125	90	88	15	____	___	____

	Selling Price		Markup Rate Based on Cost	Selling Price Equivalent Percent	Cost Price	
11.	$180	75	73%	____	____	____
12.	432	90	80	____	____	____
13.	58	25	82	____	____	____
14.	105	45	65	____	____	____
15.	117	50	85	____	____	____

Word Problems

Exercise B   Solve the following problems:

16. A furniture store paid $465.75 for a sofa. The store's markup rate is 65% of the cost price.

    (a) Find the markup amount.
    (b) Find the selling price of the sofa.

17. A TV set sells for $380.75. The cost price is $235.50.

    (a) Find the markup amount.
    (b) Find the rate of markup based on the cost.

18. A retailer marks up all his merchandise 75% of the cost price. He sells a line of dresses at $39.95 each.

    (a) Find the equivalent selling price percent.
    (b) What should be the cost price of the dresses?

Unit 5: Review of Chapter 9

Find the discount date and the due date for each of the following:

	Invoice Date	Terms	Discount Date	Due Date
1.	February 19	2/10 n 60	____	____
2.	December 5	2/10 n 90	____	____

Fill in the discount date and the due date. Then find the net amount *or* full amount, as required for each of the following invoices:

	Invoice Date	Terms	Amount of Invoice	Discount Date	Due Date	Invoice Paid	Net Amount	Full Amount
3.	8/15	2/10 n 45	$ 875 48	_____	_____	8/23	_____	_____
4.	11/17	3/15 n 60	1,673 88	_____	_____	12/1	_____	_____

Find the trade discount and the invoice price for each of the following:

	List Price	Trade Discount Percent	Trade Discount	Invoice Price
5.	$628 75	45%	_____	_____
6.	357 45	40	_____	_____

Find the invoice price for the following. (Use the direct method of subtracting from 100%.)

	List Price	Trade Discount Percent	Invoice Price
7.	$675 95	55%	_____

Find the amounts of discounts and the invoice price for each of the following:

	List Price	Trade Discount Percents	First Discount	Second Discount	Third Discount	Invoice Price
8.	$142 95	40%, 10%, 5%	_____	_____	_____	_____
9.	493 65	45%, 8%, 4%	_____	_____	_____	_____

Find the invoice price for the following:

	List Price	Trade Discount Percents	Invoice Price
10.	$237 50	50%, 5%, 5%	_____

Find the single discount percent that is equal to the series of discounts for each of the following:

	Trade Discount Percents	Single Discount Percent
11.	55%, 10%, 5%	——
12.	40%, 5%, 3%	——

Find the markup amount and the cost price on the following item:

	Selling Price	Markup Percent	Markup Amount	Cost Price
13.	$183 | 75	35%	—— | —	—— | —

Find the cost price *or* markup amount, *and* the markup percent based on the selling price on each of the following items:

	Selling Price	Cost Price	Markup Amount	Markup Percent
14.	$232 | 75	—— | —	$104 | 75	——
15.	399 | 95	—— | —	149 | 98	——

Find the equivalent cost percent and the selling price on the following item. (Markup percent is based on selling price.)

	Cost Price	Markup Percent	Equivalent Cost Percent	Selling Price
16.	$247 | 95	45%	——	—— | —

Find the markup amount (based on cost price) and the selling price on each of the following items:

	Cost Price	Markup Percent	Markup Amount	Selling Price
17.	$ 98 | 75	60%	—— | —	—— | —
18.	289 | 90	65	—— | —	—— | —

Find the markup amount and the markup rate based on the cost price on the following item:

	Cost Price		Markup Amount		Selling Price		Markup Rate
19.	$228	25	⎯⎯	⎯	$351	50	⎯⎯

20. An invoice dated 5/15 for the amount of $575.65 was paid on 6/1. If the terms were 3/15 n 30, what was the amount of the payment?

21. The list price of a TV set is $329.75 less 30%, 50%, and 10%.

    (a) Find the single discount percent.
    (b) Find the invoice price of the TV.

22. The markup amount of a dress is $23.75. If the markup based on cost is 60%, what is the selling price of the dress?

The Mathematics of Management and Finance

Unit 1: Calculating Payroll Costs

Company personnel are paid in several different ways, depending upon the size and nature of the business:

- **Salaried employees** earn an annual income, regardless of the hours expended or the output generated.
- **Hourly employees** work by the hour. Forty hours per week (8 hours per day for 5 days) is full-time work. If an hourly worker works more than 40 hours per week, he/she gets overtime pay.
- **Piecework employees** are paid by output, receiving a specified amount of money for each unit produced.
- **Commission employees** are paid a specified percent of the sales they generate. These are usually salespeople.

SALARIED EMPLOYEES

Salaried employees earn an annual wage regardless of the hours spent or the output at the office. They do not receive extra pay for extra hours worked at the office or for working at home to meet a deadline, to prepare for a meeting or presentation, or to come in weekends or holidays. Similarly, there is no loss in salary if the employee misses a reasonable number of days for sickness or to take care of family or other personal business. Most often, these employees are professionals. They have earned an advanced degree in a college, university, or other postsecondary institution or have specialized skills that are of high value to the company.

Jacinta is a salaried employee and earns an annual income of $45,600.00. What is her monthly income?

SOLUTION

Divide her salary by 12:

$$\frac{45,600}{12} = 3,800$$

Jacinta's monthly income is $3,800.00

John is a doctor and works in a lab. He earns an annual income of $125,000.00. What is his monthly income?

SOLUTION

Divide his salary by 12:

$$\frac{125,000}{12} = 10,416.67$$

John's monthly income is $10,416.67.

HOURLY EMPLOYEES

Hourly employees are paid a set rate per hour. They are then paid for the number of hours they work. They are paid on the basis of a 35–40 hour week, depending on the business. If they work *more* than the regular, accepted workweek, they are paid **overtime** wages (which are typically 1½ times the regular hourly wage). This is usually called **time and a half**. If an hourly worker is needed on weekends, holidays, or overnight, businesses might pay them at twice the regular hourly rate, or **double time**. Often, hourly workers will be paid for at least ½ hour work. That is, if they work 7 hours, 15 minutes, they will get paid for 7½ hours. Businesses rarely pay hourly workers ¼ an hourly wage. Sometimes, businesses will pay a full hour for those 15 minutes.

Maura is a line worker for the power company. She makes $28.50 per hour for a 40-hour week, and she makes time and a half for overtime. Last week, she worked 46 hours, 30 minutes. The power company pays only whole hours, so she was paid for 40 hours at the full-time rate and 7 hours at the overtime rate. How much did she make last week?

SOLUTION

First calculate the regular hours:

$$40 \times 28.50 = 1,140$$

Next calculate the overtime hours. She earns $42.75 per hour (which is 28.50 + 14.25):

$$7 \times 42.75 = 299.25$$

Now calculate her total wages by adding up the regular time and the overtime:

$$1,140 + 299.25 = 1,439.25$$

Maura earned $1,439.25 last week.

Jeanne is a troubleshooter for the telephone company. She makes $30.80 per hour for a 40-hour week, and she makes time and a half for overtime. Her contract with the phone company states that for each hour more than 8 in a single day, she will earn overtime (time and a half). Last week, she worked according to the hourly schedule below. How much did she make last week?

Monday		Tuesday		Wednesday		Thursday		Friday		Totals	
R	OT	R	OT	R	OT	R	OT	R	OT	R	OT
8	1	6	–	8	2	7	–	8	$2\frac{1}{2}$	37	$5\frac{1}{2}$

R = regular hours worked
OT = overtime

SOLUTION

First observe that Jeanne worked 37 hours full time. We calculate the regular hours:

$$37 \times 30.80 = 1,139.60$$

Now calculate the overtime hours. She earns $46.20 per hour (which is 30.80 + 15.40):

$$5\,\tfrac{1}{2} \times 46.20 = 254.10$$

Now calculate her total wages by adding up the regular time and the overtime:

$$1,139.60 + 254.10 = 1,393.70$$

Jeanne earned $1,393.70 that week.

PIECEWORK EMPLOYEES

In many businesses, employees are paid for a job or an item they produce. Employees are paid a specified amount for each article or unit they produce. The apparel industry is typical of this. Individuals who can make clothes are paid a specified amount of money for each shirt, dress, or trousers they make. The articles are called pieces, and the amount of money employees earn for producing each article is called the **piece rate**.

Ali Hassan works in piecework. He makes $1.20 per piece. One week, he made the following pieces: Monday—48, Tuesday—52, Wednesday—56, Thursday—54, and Friday—58. Find the total wages he earned that week.

SOLUTION

First add up the total pieces made that week:

$$48 + 52 + 56 + 54 + 58 = 268$$

Now calculate the wages based on 268 pieces produced:

$$268 \times 1.20 = 321.60$$

Ali earned $321.60 that week.

Bill Johns is a part-time custodian. He cleans store showrooms in town early in the morning, making $14.00 per room. One week, he cleaned the following number of rooms: Monday—7, Tuesday—5, Wednesday—6, Thursday—8, Friday—8. Find the total wages he earned that week.

SOLUTION

First add up the totals of the pieces for the week:

$$7 + 5 + 6 + 8 + 8 = 34$$

Now calculate the wages based on 34 rooms cleaned:

$$34 \times 14.00 = 476$$

Bill earned $476.00 that week.

COMMISSION EMPLOYEES

In most businesses, salespeople earn a commission from all the sales they make. A **commission** is usually a percent of the dollar value of sales. Real estate agents earn about 10–15% of the sale of a house. Sometimes the commission is a single-dollar value on each sale, and that is called a **straight commission**.

Some salespeople are on both a salary and a commission. In other words, they make a certain guaranteed weekly wage (called a **base salary**) and then a commission for each of their sales above a certain amount, called a **quota**. This allows them to be guaranteed a steady income, for their mortgage and other expenses. Then they can earn more by making sales. The base salary is usually not very high. This arrangement is called a **salary plus commission basis**.

EXAMPLE 7

Anastasia is a real estate salesperson. She makes 15% on each house she sells. One month, she sold a house worth $56,800, another house worth $89,200, and a third house worth $49,000. Find the total commission she earned that month.

SOLUTION

First add up the value of the houses:

$$56,800 + 89,200 + 49,000 = 195,000$$

Now calculate her commission based on 15%:

$$195,000 \times 0.15 = 29,250$$

Anastasia earned $29,250.00 that month.

EXAMPLE 8

Ungumi sells large appliances. He earns a salary of $150 per week plus 10% on each appliance he sells over his quota of $2,000. One week, he sold $6,750. Find his total earnings that week.

SOLUTION

First, he gets $150 per week.

Now calculate his commission based on 10% over $2,000:

$$6,750 - 2,000 = 4,750$$
$$4,750 \times 0.10 = 475$$

Now calculate his total earnings:

$$150 + 475 = 625$$

Ungumi earned $625.00 that week.

EXERCISES

Exercise A   Complete the following partial payroll record based on a 40-hour week, with time and a half for overtime:

TRIANGLE MANUFACTURING COMPANY

Payroll for Week Ending April 8 20—

	Employee	M	T	W	Th	F	Regular Hours	Regular Rate	Total Regular Amount	Overtime Hours	Overtime Rate	Total Overtime Amount	Gross Pay
1.	A	9	$8\frac{1}{2}$	6	$10\frac{1}{2}$	9	___	$5 75	___ ___	___	___	___ ___	___ ___
2.	B	8	7	$10\frac{1}{2}$	$9\frac{1}{2}$	6	___	4 50	___ ___	___	___	___ ___	___ ___

TRIANGLE MANUFACTURING COMPANY

Payroll for Week Ending April 8 20—

	Employee	M	T	W	Th	F	Regular Hours	Regular Rate		Total Regular Amount	Overtime Hours	Overtime Rate	Total Overtime Amount	Gross Pay
3.	C	$9\frac{1}{4}$	$10\frac{1}{2}$	$8\frac{1}{4}$	$9\frac{1}{4}$	10	—	4	25	—	—	—	—	—
4.	D	6	10	$8\frac{1}{2}$	$9\frac{1}{4}$	9	—	6	20	—	—	—	—	—
5.	E	$9\frac{1}{4}$	$10\frac{1}{2}$	9	$8\frac{1}{4}$	9	—	5	00	—	—	—	—	—

Complete the following partial payroll record, with overtime paid for hours worked beyond 8 hours each day:

DURABLE PICTURE FRAME COMPANY

Payroll for Week Ending April 15 20—

	Employee	M	T	W	Th	F	Regular Hours	Regular Rate		Total Regular Amount	Overtime Hours	Overtime Rate	Total Overtime Amount	Gross Pay
6.	A	9	6	$8\frac{1}{2}$	7	8	—	$5	35	—	—	—	—	. —
7.	B	10	8	7	9	$8\frac{1}{2}$	—	7	80	—	—	—	—	. —
8.	C	8	$9\frac{1}{2}$	$10\frac{1}{2}$	7	6	—	5	20	—	—	—	—	. —
9.	D	8	$10\frac{1}{2}$	6	$9\frac{1}{2}$	9	—	7	80	—	—	—	—	. —
10.	E	9	6	7	$10\frac{1}{2}$	12	—	4	75	—	—	—	—	. —

Exercise B Using a calculator, find the total wages for each employee. Add horizontally and use equivalent parts of $1 where possible. Multiply by the decimal value of the rate or the equivalent fraction.

	Employee	Number of Pieces					Total Pieces	Piece Rate	Total Wages
		M	T	W	Th	F			
11.	Allen, J.	105	115	125	118	126	—	$0.40	— —
12.	Bauer, S.	119	121	116	124	118	—	$0.33\frac{1}{3}$ $(\frac{1}{3})$	— —
13.	Berg, M.	100	112	117	114	109	—	0.45	— —
14.	Diaz, J.	118	122	104	123	113	—	0.35	— —
15.	Wilson, J.	118	121	117	114	119	—	0.75	— —

Find the total wages for each employee.

	Employee	Number of Dozens					Total Dozens	Piece Rate		Total Wages	
		M	T	W	Th	F					
16.	Adams, C.	13	$11\frac{1}{2}$	$12\frac{1}{4}$	$10\frac{1}{2}$	12	_____	$2	95	_____	_____
17.	Bond, J.	15	$12\frac{1}{4}$	$14\frac{1}{2}$	13	$15\frac{1}{2}$	_____	2	78	_____	_____
18.	Crane, V.	$13\frac{1}{2}$	15	$12\frac{1}{4}$	$11\frac{1}{2}$	14	_____	2	92	_____	_____
19.	Dean, J.	14	$13\frac{1}{4}$	$15\frac{1}{2}$	13	12	_____	3	05	_____	_____
20.	Watts, M.	$14\frac{1}{4}$	$13\frac{1}{2}$	15	12	$13\frac{1}{4}$	_____	2	92	_____	_____

Find the total earnings for each of the following salespersons:

	Salesperson	Salary	Weekly Sales		Sales Quota	Net Sales		Commission	Total Earnings	
21.	Edwards, G.	$85	$ 2,185	60	None	_____	_____	13%	_____	_____
22.	Ehlers, M.	None	6,524	00	None	_____	_____	5	_____	_____
23.	Farrel, S.	$75	3,478	50	None	_____	_____	8	_____	_____
24.	Lopez, J.	$95	7,460	00	$2,000	_____	_____	4	_____	_____
25.	Lucas, B.	None	18,260	00	$5,500	_____	_____	3	_____	_____

Word Problems

Exercise C   Solve the following problems:

26. Richie worked the following hours for the week ending August 15: 9, $9\frac{1}{2}$, 10, $8\frac{3}{4}$, and 9 hours. His hourly rate is $6.75, with time and a half for hours worked beyond 40 hours. What was his gross pay for the week?

27. Marvin works a 35-hour week, with time and a half for hours worked over 35 hours and double time for hours worked on Saturday. Last week he worked $8\frac{1}{2}$ hours on Monday, $7\frac{1}{2}$ on Tuesday, 9 on Wednesday, 8 on Thursday, $9\frac{1}{2}$ on Friday, and 4 on Saturday. If his hourly rate is $7.65, how much was his gross pay?

28. Albert's hourly rate is $6.75, with time and a half for hours worked beyond 8 hours a day. Last week he worked the following hours: Monday, $8\frac{1}{2}$; Tuesday, $9\frac{3}{4}$; Wednesday, 8; and Thursday, 10. Friday was a legal holiday with pay. Find his gross pay for the week.

29. An office employs six clerks earning the following weekly salaries: $285, $315, $265, $247, $348, and $465. What is the total yearly payroll for all the clerks?

30. Mary is a sewing machine operator and is paid on a piecework basis. For the week ending July 8, she completed the following number of blouses: Monday, 215; Tuesday, 235; Wednesday, 225; Thursday, 240; and Friday, 228. If she is paid 45¢ for each completed blouse, how much did she earn for the week?

31. Jean is paid 75¢ for every toaster she assembles. Last week she assembled the following numbers of toasters: 58, 53, 61, 57, and 56. Find her earnings for the week.

32. Sarah is paid $5\frac{1}{2}$% commission on sales. Last week her sales for each day were as follows: $345.75, $298.50, $363.85, $363.80, and $375.90. How much commission did she earn for the week?

33. Greta earns $65 on each vacuum cleaner she sells. If the price of the vacuum cleaner is $285.75, find the rate of commission.

Unit 2: Calculating Payroll Deductions

All businesses are required to make tax deductions from employees' paychecks. Employees do their own taxes, of course. So these deductions are based on the employees' estimated tax bill for the year. These deductions also include FICA (Social Security and medicare), state, and local taxes. In addition, deductions may be made for health benefits, savings bonds, credit unions, and/or individual retirement accounts.

FEDERAL INCOME TAX

An employee's federal income tax withholdings are based on the gross earnings and the number of exemptions he/she claims. For example, a male employee may claim one exemption for himself, one for his wife, and one for each of his dependent children. The amount of tax withheld will decrease with each exemption. A single male with no dependents, on the other hand, will be able to claim only one exemption: himself.

At the end of the year and before the tax day of April 15, the taxpayer files a yearly income tax return to determine his or her income tax. If the income tax withheld during the year is more than the tax owed, then a **refund** will be due to the taxpayer. If the income tax withheld during the year is less than the tax owed, the taxpayer will be required to send the Internal Revenue Service the difference.

To estimate the amount of federal withholding tax, employers consult federal tax tables, such as Tables 1 and 2 on pages 255 and 256.

NOTE: For the exercises in this chapter, use the figures from these tables, as such figures have changed and will continue to change periodically.

TABLE 1
FEDERAL INCOME TAX WITHHOLDING SCHEDULE

SINGLE Persons—**WEEKLY** Payroll Period

(For Wages Paid in 2000)

If the wages are—		And the number of withholding allowances claimed is—										
At least	But less than	0	1	2	3	4	5	6	7	8	9	10
		The amount of income tax to be withheld is—										
$600	$610	92	77	67	59	51	43	35	27	18	10	2
610	620	95	80	68	60	52	44	36	28	20	12	4
620	630	98	83	70	62	54	46	38	30	21	13	5
630	640	101	85	71	63	55	47	39	31	23	15	7
640	650	103	88	73	65	57	49	41	33	24	16	8
650	660	106	91	76	66	58	50	42	34	26	18	10
660	670	109	94	79	68	60	52	44	36	27	19	11
670	680	112	97	82	69	61	53	45	37	29	21	13
680	690	115	99	84	71	63	55	47	39	30	22	14
690	700	117	102	87	72	64	56	48	40	32	24	16
700	710	120	105	90	75	66	58	50	42	33	25	17
710	720	123	108	93	78	67	59	51	43	35	27	19
720	730	126	111	96	81	69	61	53	45	36	28	20
730	740	129	113	98	83	70	62	54	46	38	30	22
740	750	131	116	101	86	72	64	56	48	39	31	23
750	760	134	119	104	89	74	65	57	49	41	33	25
760	770	137	122	107	92	77	67	59	51	42	34	26
770	780	140	125	110	95	79	68	60	52	44	36	28
780	790	143	127	112	97	82	70	62	54	45	37	29
790	800	145	130	115	100	85	71	63	55	47	39	31
800	810	148	133	118	103	88	73	65	57	48	40	32
810	820	151	136	121	106	91	76	66	58	50	42	34
820	830	154	139	124	109	93	78	68	60	51	43	35
830	840	157	141	126	111	96	81	69	61	53	45	37
840	850	159	144	129	114	99	84	71	63	54	46	38
850	860	162	147	132	117	102	87	72	64	56	48	40
860	870	165	150	135	120	105	90	74	66	57	49	41
870	880	168	153	138	123	107	92	77	67	59	51	43
880	890	171	155	140	125	110	95	80	69	60	52	44
890	900	173	158	143	128	113	98	83	70	62	54	46
900	910	176	161	146	131	116	101	86	72	63	55	47
910	920	179	164	149	134	119	104	88	73	65	57	49
920	930	182	167	152	137	121	106	91	76	66	58	50
930	940	185	169	154	139	124	109	94	79	68	60	52
940	950	187	172	157	142	127	112	97	82	69	61	53
950	960	190	175	160	145	130	115	100	85	71	63	55
960	970	193	178	163	148	133	118	102	87	72	64	56
970	980	196	181	166	151	135	120	105	90	75	66	58
980	990	199	183	168	153	138	123	108	93	78	67	59
990	1,000	201	186	171	156	141	126	111	96	81	69	61
1,000	1,010	204	189	174	159	144	129	114	99	84	70	62
1,010	1,020	207	192	177	162	147	132	116	101	86	72	64
1,020	1,030	210	195	180	165	149	134	119	104	89	74	65
1,030	1,040	213	197	182	167	152	137	122	107	92	77	67
1,040	1,050	215	200	185	170	155	140	125	110	95	80	68
1,050	1,060	218	203	188	173	158	143	128	113	98	82	70
1,060	1,070	221	206	191	176	161	146	130	115	100	85	71
1,070	1,080	224	209	194	179	163	148	133	118	103	88	73
1,080	1,090	227	211	196	181	166	151	136	121	106	91	76
1,090	1,100	229	214	199	184	169	154	139	124	109	94	79
1,100	1,110	232	217	202	187	172	157	142	127	112	96	81
1,110	1,120	235	220	205	190	175	160	144	129	114	99	84
1,120	1,130	238	223	208	193	177	162	147	132	117	102	87
1,130	1,140	241	225	210	195	180	165	150	135	120	105	90
1,140	1,150	243	228	213	198	183	168	153	138	123	108	93
1,150	1,160	246	231	216	201	186	171	156	141	126	110	95
1,160	1,170	249	234	219	204	189	174	158	143	128	113	98
1,170	1,180	252	237	222	207	191	176	161	146	131	116	101
1,180	1,190	256	239	224	209	194	179	164	149	134	119	104
1,190	1,200	259	242	227	212	197	182	167	152	137	122	107
1,200	1,210	262	245	230	215	200	185	170	155	140	124	109
1,210	1,220	265	248	233	218	203	188	172	157	142	127	112
1,220	1,230	268	251	236	221	205	190	175	160	145	130	115
1,230	1,240	271	254	238	223	208	193	178	163	148	133	118
1,240	1,250	274	257	241	226	211	196	181	166	151	136	121

$1,250 and over   Use Table 1(a) for a **SINGLE person** on page 34. Also see the instructions on page 32.

TABLE 2
FEDERAL INCOME TAX WITHHOLDING SCHEDULE

MARRIED Persons—**WEEKLY** Payroll Period
(For Wages Paid in 2000)

If the wages are—		And the number of withholding allowances claimed is—										
At least	But less than	0	1	2	3	4	5	6	7	8	9	10
		The amount of income tax to be withheld is—										
$740	$750	93	85	77	69	61	53	45	37	29	20	12
750	760	95	87	78	70	62	54	46	38	30	22	14
760	770	96	88	80	72	64	56	48	40	32	23	15
770	780	98	90	81	73	65	57	49	41	33	25	17
780	790	99	91	83	75	67	59	51	43	35	26	18
790	800	101	93	84	76	68	60	52	44	36	28	20
800	810	102	94	86	78	70	62	54	46	38	29	21
810	820	104	96	87	79	71	63	55	47	39	31	23
820	830	105	97	89	81	73	65	57	49	41	32	24
830	840	107	99	90	82	74	66	58	50	42	34	26
840	850	108	100	92	84	76	68	60	52	44	35	27
850	860	110	102	93	85	77	69	61	53	45	37	29
860	870	111	103	95	87	79	71	63	55	47	38	30
870	880	113	105	96	88	80	72	64	56	48	40	32
880	890	114	106	98	90	82	74	66	58	50	41	33
890	900	116	108	99	91	83	75	67	59	51	43	35
900	910	117	109	101	93	85	77	69	61	53	44	36
910	920	119	111	102	94	86	78	70	62	54	46	38
920	930	120	112	104	96	88	80	72	64	56	47	39
930	940	122	114	105	97	89	81	73	65	57	49	41
940	950	125	115	107	99	91	83	75	67	59	50	42
950	960	128	117	108	100	92	84	76	68	60	52	44
960	970	131	118	110	102	94	86	78	70	62	53	45
970	980	133	120	111	103	95	87	79	71	63	55	47
980	990	136	121	113	105	97	89	81	73	65	56	48
990	1,000	139	124	114	106	98	90	82	74	66	58	50
1,000	1,010	142	127	116	108	100	92	84	76	68	59	51
1,010	1,020	145	130	117	109	101	93	85	77	69	61	53
1,020	1,030	147	132	119	111	103	95	87	79	71	62	54
1,030	1,040	150	135	120	112	104	96	88	80	72	64	56
1,040	1,050	153	138	123	114	106	98	90	82	74	65	57
1,050	1,060	156	141	126	115	107	99	91	83	75	67	59
1,060	1,070	159	144	128	117	109	101	93	85	77	68	60
1,070	1,080	161	146	131	118	110	102	94	86	78	70	62
1,080	1,090	164	149	134	120	112	104	96	88	80	71	63
1,090	1,100	167	152	137	122	113	105	97	89	81	73	65
1,100	1,110	170	155	140	125	115	107	99	91	83	74	66
1,110	1,120	173	158	142	127	116	108	100	92	84	76	68
1,120	1,130	175	160	145	130	118	110	102	94	86	77	69
1,130	1,140	178	163	148	133	119	111	103	95	87	79	71
1,140	1,150	181	166	151	136	121	113	105	97	89	80	72
1,150	1,160	184	169	154	139	123	114	106	98	90	82	74
1,160	1,170	187	172	156	141	126	116	108	100	92	83	75
1,170	1,180	189	174	159	144	129	117	109	101	93	85	77
1,180	1,190	192	177	162	147	132	119	111	103	95	86	78
1,190	1,200	195	180	165	150	135	120	112	104	96	88	80
1,200	1,210	198	183	168	153	137	122	114	106	98	89	81
1,210	1,220	201	186	170	155	140	125	115	107	99	91	83
1,220	1,230	203	188	173	158	143	128	117	109	101	92	84
1,230	1,240	206	191	176	161	146	131	118	110	102	94	86
1,240	1,250	209	194	179	164	149	134	120	112	104	95	87
1,250	1,260	212	197	182	167	151	136	121	113	105	97	89
1,260	1,270	215	200	184	169	154	139	124	115	107	98	90
1,270	1,280	217	202	187	172	157	142	127	116	108	100	92
1,280	1,290	220	205	190	175	160	145	130	118	110	101	93
1,290	1,300	223	208	193	178	163	148	133	119	111	103	95
1,300	1,310	226	211	196	181	165	150	135	121	113	104	96
1,310	1,320	229	214	198	183	168	153	138	123	114	106	98
1,320	1,330	231	216	201	186	171	156	141	126	116	107	99
1,330	1,340	234	219	204	189	174	159	144	129	117	109	101
1,340	1,350	237	222	207	192	177	162	147	131	119	110	102
1,350	1,360	240	225	210	195	179	164	149	134	120	112	104
1,360	1,370	243	228	212	197	182	167	152	137	122	113	105
1,370	1,380	245	230	215	200	185	170	155	140	125	115	107
1,380	1,390	248	233	218	203	188	173	158	143	128	116	108

$1,390 and over   Use Table 1(b) for a **MARRIED person** on page 34. Also see the instructions on page 32.

A single person earning $655 per week with zero (0) exemptions would have $106 withheld from his/her paycheck.

Question: Why would a person claim 0 exemptions? Wouldn't that person want to at least claim himself/herself?

Answer: A person may want more money withheld each pay period to ensure that he/she receives a refund at tax time.

 A married person earning $1,255 per week with three (3) exemptions would have $169 withheld from his/her paycheck.

SOCIAL SECURITY AND MEDICARE (FICA) TAXES

The other federal tax deducted from weekly income under the Federal Insurance Contributions Act is the FICA tax. Commonly known as the Social Security tax, this tax provides retirement income, disability, and survivor's benefits. It also includes the Medicare program, which provides partial benefits on medical expenses for people age 65 or over.

For 2000, the rate of the social security tax was 6.20% on the first $76,200.00 of income, or a maximum of $4,724.40 No social security tax is paid on income past the first $76,200.00. The rate of the medicare tax was 1.45% with no wage limit. The combined FICA tax, therefore, was 7.65%.

Under this act, the employer must contribute an amount equal to the amount contributed by the employee, which is credited to that employee in the Social Security Administration under his/her social security number.

If an employee earned more than $76,200.00, the FICA deduction stopped on the week that his/her deductions equalled the maximum ($4,724.40). On the following week, the deductions then increased by the amount of the FICA deduction.

 Escobar earns $1,650 per week, or $85,800.00 per year. How much money in FICA taxes are deducted each week?

SOLUTION

To solve this, multiply the combined FICA tax (7.65% or 0.0765) by the weekly wage:

$$x = 1,650 \times 0.0765$$
$$x = 126.225$$

The amount deducted each week is $126.23.

EXAMPLE 4 Using the information in Example 3, on which week of the year will the FICA tax deduction stop?

SOLUTION

To find the week, divide the maximum salary ($76,200.00) by the weekly wages ($1,650.00). This will give the week that the FICA deductions will stop:

$$\frac{7,6200}{1,650} = 46.18$$

FICA deductions will end after the 46th week of the year.

EXERCISES

Exercise A Find the FICA tax deduction for each of the following weekly wages. The rate is 7.65%.

	Weekly Salary		FICA Tax	
1.	$475	50	⎯	⎯
2.	453	45	⎯	⎯
3.	463	70	⎯	⎯
4.	385	75	⎯	⎯
5.	575	80	⎯	⎯

Find the week in which no FICA tax will be deducted from the following weekly salaries:

	Weekly Salary	Week
6.	$ 985	⎯
7.	1,005	⎯
8.	1,670	⎯
9.	2,000	⎯
10.	2,400	⎯

Exercise B   Complete the following table, using the federal income tax tables on pages 255–256. Remember that the FICA tax rate is 7.65%.

SANDS ELECTRONIC COMPANY

Week Ending ___July 10___ 20—

	Employee	Number of Exemptions	Weekly Salary		Income Tax		FICA Tax		Take-Home Pay	
11.	A: married	3	$767	50						
12.	B: married	4	793	75						
13.	C: single	1	785	50						
14.	D: married	6	873	45						
15.	E: single	4	680	75						

Word Problems

Exercise C   Use the tax tables on pages 255–256 to solve the following problems:

16. Sandra is single and earns $685.50 a week. If she claims no dependents, what will be her take-home pay?

17. Henry earns $66,500 a year. He is married and claims three dependents. How much will his take-home pay be?

18. William earns $880 a week. He is married and claims five exemptions. How much will his net pay be?

Unit 3: Review of Chapter 10

SUMMARY OF KEY POINTS

- *Regular work week*:   35- or 40-hour work week.
- *Overtime hours*:   Hours worked beyond a 35- or 40-hour week, or beyond an 8-hour day.
- *Overtime rate*:   Two times or one and a half times the regular rate.
- *Piecework employee*:   Employee paid a specified amount for each unit produced.
- *Commission employee*:   Employee paid a specified percentage of sales.

Complete the following partial payroll record based on a 40-hour week, with time and a half for overtime:

	Employee	M	T	W	Th	F	Hourly Rate		Regular Pay		Overtime Pay		Gross Pay	
1.	A	$9\frac{1}{2}$	8	$10\frac{3}{4}$	$9\frac{1}{4}$	$8\frac{1}{2}$	$7	75	——	—	——	—	——	—
2.	B	9	$10\frac{3}{4}$	$12\frac{1}{2}$	8	$11\frac{3}{4}$	6	85	——	—	——	—	——	—

Complete the following partial payroll record, with overtime paid for hours worked beyond 8 hours each day:

	Employee	M	T	W	Th	F	Hourly Rate		Regular Pay		Overtime Pay		Gross Pay	
3.	A	9	$8\frac{3}{4}$	$10\frac{1}{2}$	$9\frac{1}{2}$	10	$7	65	——	——	——	——	——	——

Find the total wages for each employee in Problems 4 through 6.

		Number of Pieces					Total Pieces	Piece Rate	Total Wages	
	Employee	M	T	W	Th	F				
4.	A	85	92	89	98	95	——	$0.60	——	—
5.	B	225	230	227	236	228	——	0.35	——	—

		Number of Dozens					Total Dozens	Piece Rate	Total Wages	
	Employee	M	T	W	Th	F				
6.	A	23	26	27	24	25	——	$2	53	——	—

Find the FICA tax deductions for each of the following weekly wages. The rate is 7.65%.

	Weekly Salary		FICA Tax	
7.	$463	85	——	——
8.	568	70	——	——

Find the week in which no FICA tax will be deducted from the following weekly salaries:

	Weekly Salary		Week
9.	$1,538	46	_____
10.	1,778	85	_____

Using the income tax tables on pages 255–256, complete the following partial payroll. The FICA tax rate is 7.65%.

	Employee	Exemptions	Weekly Salary		Income Tax		FICA Tax		Take-Home Pay	
11.	A: married	4	$763	37	_____	___	_____	___	_____	___
12.	B: married	5	897	42	_____	___	_____	___	_____	___

Solve the following problems:

13. Victor's total earnings last week were $468.50. If his sales were $6,790.90 and his rate of commission is $5\frac{1}{2}\%$, find his base salary.

14. Mary Johnson earns a base salary of $75 plus a commission of $12\frac{1}{2}\%$ on sales over $1,500. Last week her total sales were $3,967.80. How much was her total salary for the week?

Basic Bookkeeping

Unit 1: The Balance Sheet

Bookkeeping in business is both very simple and very difficult. It is simple because only addition and subtraction skills are needed. It is difficult because getting accurate numbers for the various entries on the sheets and statements can be hard to do.

The balance sheet is the basic statement of the worth of a company. The balance sheet equation can be written in two ways:

$$Assets = Liabilities + Equity$$
$$Equity\ (net\ worth) = Assets - Liabilities$$

- **Assets** are all resources of the company: cash, accounts receivable, inventory of goods, supplies, and all other liquid holdings of the company, including buildings, office supplies, land, and vehicles.
- **Liabilities** are all debits, bills, and other obligations the company will have to pay out.
- **Equity** is the net value of the company, that is, the assets minus the liabilities.

Mr. Pat Walsh has a security business. At the end of the year, he prepared a balance sheet to show his net worth. His assets are as follows. He has $12,567 cash on hand and $7,895 accounts receivable. He bought his building at $250,000, and he has a $100,000 mortgage on it. He has 5 vehicles, with a total worth of $75,000. His furniture is worth $5,620, and the office supplies are worth $500. His liabilities are as follows. His payroll is $16,000, he has utilities of $4,520, and his accounts payable are $12,580. What does his prepared balance sheet look like?

SOLUTION

Assets:			
Cash on hand:	$	12,567	
Accounts receivable:	$	7,895	
Building:	$	250,000	
Vehicles:	$	75,000	
Furniture:	$	5,620	
Office supplies:	$	500	
Total:	$	351,582	
Liabilities:			
Mortgage:	$	100,000	
Payroll:	$	16,000	
Utilities:	$	4,520	
Accounts payable:	$	12,580	
Total:	$	133,100	
Equity (net worth):	$351,582 – $133,100 = $218,482		

EXAMPLE 2 The Scrantom's Stationery Company has assets as follows. It has $7,556 cash on hand and $10,895 accounts receivable. The company has 2 vehicles, with a total worth of $25,000. It has furniture worth $14,750, inventory worth $64,000, and office supplies worth $853. Liabilities are as follows. The business rents space in a building, and its rent is $2,300 per month. It owes $12,580 on the two vehicles. The payroll for the business is $33,780, the accounts payable is $9,837, and overhead is $9,580. What does the company's prepared balance sheet look like?

SOLUTION

Assets:			
Cash on hand:	$	7,556	
Accounts receivable:	$	10,895	
Vehicles:	$	25,000	
Furniture:	$	14,750	
Inventory:	$	64,000	
Office supplies:	$	853	
Total:	$	123,054	

Liabilities:			
Rent:	$ 2,300	
Vehicle debt:	$ 12,580	
Payroll:	$ 33,780	
Accounts Payable:	$ 9,837	
Overhead:	$ 9,580	
Total:	$ 68,077	
Equity (net worth):	$123,054 − $68,077 = $54,977	

ELECTRONIC SPREAD SHEET

An electronic spreadsheet makes doing these calculations very easy. You can set up a spreadsheet that has all these entries, including the titles and the numbers. You can have the spreadsheet calculate the values for you. Setting up a spreadsheet will allow you to calculate numerous problems by simply changing the numbers and titles in them. Here is the Scrantom's Stationery Company's balance sheet set up on a spreadsheet for your consideration:

Scrantom's Balance Sheet		
Assets:		
Cash on hand:	$ 7,556.00	
Accounts receivable:	$ 10,895.00	
Vehicles:	$ 25,000.00	
Furniture:	$ 14,750.00	
Inventory:	$ 64,000.00	
Office supplies:	$ 853.00	
Total assets:	$ 123,054.00	
Liabilities:		
Rent:	$ 2,300.00	
Vehicle debt:	$ 12,580.00	
Payroll:	$ 33,780.00	
Accounts payable:	$ 9,837.00	
Overhead:	$ 9,580.00	
Total liabilities:	$ 68,077.00	
Equity (net worth):	$123,054 − $68,077 = $54,977	

EXERCISES

Exercise A   Using these figures, find the total assets, the total liabilities, and the net worth of each company.

1.

	Frear & Co. Balance Sheet		
Assets:			
	Cash on hand:	$ 5,342.00	
	Accounts receivable:	$ 23,350.00	
	Vehicles:	$ 32,750.00	
	Furniture:	$ 9,452.00	
	Inventory:	$ 52,000.00	
	Office supplies:	$ 630.00	
	Total assets:	$	
Liabilities:			
	Rent:	$ 3,200.00	
	Vehicles debt:	$ 22,487.00	
	Payroll:	$ 12,455.00	
	Accounts payable:	$ 12,643.00	
	Overhead:	$ 10,785.00	
	Total liabilities:	$	
	Equity (net worth):	$	

2.

		RYC Balance Sheet		
Assets:				
	Cash on hand:		$ 12,955.00	
	Accounts receivable:		$ 10,895.00	
	Vehicles:		$ 56,500.00	
	Furniture:		$ 24,552.00	
	Inventory:		$ 14,000.00	
	Office supplies:		$ 767.00	
	Total assets:		$	
Liabilities:				
	Rent:		$ 3,400.00	
	Vehicles debt:		$ 32,450.00	
	Payroll:		$ 12,780.00	
	Accounts payable:		$ 8,432.00	
	Overhead:		$ 11,776.00	
	Total liabilities:		$	
	Equity (net worth):		$	

Word Problems

Exercise B   Set up a balance sheet for each of these companies. Find the total assets, liabilities, and the net worth of each.

1. At the end of the year, Flying Sparks Electronic Company calculated its assets and liabilities. It has $11,885 cash on hand and $4,562 accounts receivable. It owns its building, which is worth $450,000. The company has 1 van, with a total worth of $30,000. It has furniture worth $8,500, inventory worth $73,650, and office supplies worth $6,420. Liabilities are as follows. The mortgage on the building is $250,000, and the company owes 19,250 on the van. The payroll for the business is $30,265, and the accounts payable is $7,037. Utilities and other overhead charges are $7,865. Prepare a balance sheet for the Flying Sparks Electronic Company.

2. On December 31, Mr. Paul Blatz, owner of DaVinci's Pizza and Pasta, prepared a balance sheet. The restaurant has $10,556 cash on hand, with $5,950 accounts receivable. The company has 1 van, with a total worth of $31,500. It has ovens, pasta machines, and other furniture worth $24,750, inventory worth $4,250, and office supplies worth $150. Liabilities are as follows. The business rents space in a building, and its rent is $1,500 per month. The company owes $13,000 on the van. The payroll for the business is $28,300, the accounts payable are $7,537, and overhead (including utilities) is $8,775. What does DaVinci's Pizza and Pasta balance sheet look like?

3. Just after Christmas, Time to ReTire Auto Parts prepared its balance sheet for the new year. It has $11,236 cash on hand and $12,824 accounts receivable. The business owns its building. The business paid off the mortgage 5 years ago. The building is worth $450,000. The company has 2 cars and 2 vans, with a total worth of $80,000. It has furniture worth $18,260, inventory worth $104,450, and office supplies worth $510. Liabilities are as follows. The company owes $45,000 on the 4 vehicles. The payroll for the business is $29,520, the accounts payable are $18,500, and the overhead is $7,180. Show Time to ReTire Auto Parts' balance sheet for this year.

4. In preparation for December 31, The Big Drip Plumbing Corp. prepared its balance sheet. It has $3,225 cash on hand and $5,415 accounts receivable. The company has 3 vans, with a total worth of $85,000. It has furniture worth $3,955, inventory worth $23,050, and office supplies worth $721. Liabilities are as follows. The business rents space in a building, and its rent is $2,750 per month. The company owes $41,500 on the 3 vans. The payroll for the business is $17,180, the accounts payable are $3,750, and utilities and other overhead charges is $2,160. What does The Big Drip Plumbing Corp.'s balance sheet look like?

5. For the New Year, Janet Peterson, owner of White Glove Maid Service, prepared a balance sheet to show her net worth. Her assets are as follows. She has $10,621 cash on hand and $4,370 accounts receivable. Her inventory (vacuums, rug shampooers, and janitorial supplies) is worth $17,360. She bought her building for $350,000. She has 6 vehicles (4 vans and 2 cars), with a total worth of $105,000. Her furniture is worth $2,620, and office supplies are worth $670. Her liabilities are as follows. She has a $200,000 mortgage on her building, and she owes $65,000 on her vehicles. Her payroll is $19,500. She has utilities and other overhead charges of $8,530. Prepare a balance sheet for White Glove Maid Service.

Unit 2: The Income Statement

The balance sheet expresses the basic worth of a company. In contrast, the income statement shows the difference between the money paid out and the money taken in at a particular time. The equation used is :

$$\text{Net profit} = \text{Income} - \text{Costs (overhead)}$$

*   **Net profit** is the total income less the overhead.
*   **Income** is the total of all monies coming in, whether for goods sold or services provided.
*   **Costs** are the amount of money being paid out of the business at any one time.

The income statement could be prepared for a month, a quarter, or a year.

 Mr. Jini Kim is a private investigator. In June, he prepared an income statement. He had income of $4,500 (for investigative services) and $2,400 (for surveillance services). He paid $2,200 for rent of his office, $150 in insurance, $350 in utilities, and $450 in incidentals (advertising, hotel rooms, and so on). What does his income statement look like?

SOLUTION

Revenues:				
	Investigations:		$ 4,500	
	Surveillance:		$ 2,400	
	Total revenues:		$ 6,900	
Expenses:				
	Rent:		$ 2,200	
	Insurance:		$ 150	
	Utilities:		$ 350	
	Incidentals:		$ 450	
	Total expenses:		$ 3,150	
	Net profit = $6,900 − $3,150 = $3,750			

Mrs. Lola Quince owns a 15-unit apartment building. For the fourth quarter, she wrote up an income statement. She had rental income of $18,500. She paid $9,200 mortgage on the building, $1,506 in insurance, $1,050 in utilities, and $930 in building repairs. What does her income statement look like?

SOLUTION

Revenues:				
	Rental income:		$ 18,500	
	Total revenues:		$ 18,500	
Expenses:				
	Mortgage:		$ 9,200	
	Insurance:		$ 1,506	
	Utilities:		$ 1,057	
	Building repairs:		$ 930	
	Total expenses:		$ 12,693	
	Net profit = $18,500 − $12,693 = $5,807			

Mike Peebles runs the Soda Pop Shop. He wrote up an income statement for last month. He pays $2,520 rent on his shop. He paid $243 in insurance, $737 in utilities, $3,250 in wages to his workers, and $1,052 for supplies (ice cream, soda, hamburgers, and so on). His gross receipts last month were $9,546. What does the Soda Pop Shop's income statement look like?

SOLUTION

Revenues:				
	Gross receipts:		$ 9,546	
	Total revenues:		$ 9,546	

Expenses:				
	Rent:		$ 2,520	
	Insurance:		$ 243	
	Utilities:		$ 737	
	Wages:		$ 3,250	
	Supplies:		$ 1,052	
	Total expenses:		$ 7,802	
	Net profit = $9,546 − $7,802 = $1,744			

ELECTRONIC SPREADSHEET

As with the balance sheet, an electronic spreadsheet makes doing these calculations very easy. You can set up a spreadsheet that has all these entries, including the titles and the numbers. You can have the spreadsheet calculate the values for you. Setting up a spreadsheet will allow you to calculate numerous problems by simply changing the numbers and titles in them. Here is the Soda Pop Shop's income statement is set up on a spreadsheet:

	Soda Pop Shop Income Statement			
Revenues:				
	Gross receipts:		$ 9,546.00	
	Total revenues:		$ 9,546.00	
Expenses:				
	Rent:		$ 2,520.00	
	Vehicle debt:		$ 243.00	
	Payroll:		$ 737.00	
	Accounts payable:		$ 3,250.00	
	Overhead:		$ 1,052.00	
	Total Liabilities:		$ 7,802.00	
	Net profit:		$9,546 − $7,802 = $1,744	

EXERCISES

Exercise A Using these figures, find the total revenues, the total liabilities, and the net profit of each company. Note that sometimes a company has a negative net profit.

1.

	Lester Dutcher, Inc. Income Statement		
Revenues:			
	Gross receipts:	$ 15,250.00	
	Dividend income:	$ 1,576.45	
	Total revenues:	$	
Expenses:			
	Rent:	$ 1,455.00	
	Vehicle debt:	$ 10,650.00	
	Payroll:	$ 3,420.00	
	Accounts payable:	$ 2,555.00	
	Overhead:	$ 952.00	
	Total liabilities:	$	
	Net profit:	$	

2.

	Waring Investments, Inc. Income Statement		
Revenues:			
	Gross receipts:	$ 10,546.00	
	Investments income:	$ 5,166.75	
	Total revenues:	$	

Expenses:			
Rent:	$	1,322.00	
Vehicle debt:	$	278.00	
Payroll:	$	1,274.00	
Utilities:	$	3,537.00	
	$	820.00	
Total liabilities:	$		
Net profit:	$		

3.

Stutson St. Apts. Income Statement			
Revenues:			
Rental income:	$	15,875.00	
Total revenues:	$		
Expenses:			
Mortgage:	$	2,520.00	
Equipment debt:	$	3,445.00	
Payroll:	$	1,595.00	
Building repairs:	$	2,250.00	
Overhead (utilities & insurance):	$	3,472.00	
Total liabilities:	$		
Net profit:	$		

Word Problems

Exercise B   Set up an income statement for each of these companies. Find the total revenues, liabilities, and the net profit of each.

1. Ron McDowell has a dry cleaning business. He wrote an income statement for last month. He pays $4,190 rent on his shop. He paid $475 in insurance, $1,298 in utilities, $3,100 for wages, and $3,450 for supplies (cleaning solution, hangers, and so on). Last month, he took in $13,576. What does his income statement look like?

2. Jim Roddick runs the Hank Hoagies Sandwich Store. His quarterly receipts were as follows. He sold $7,541 in sandwiches and drinks. He pays $2,000 in rent, $532 in insurance, $798 in utilities, and $2,106 in wages. He spent $2,345 for supplies (bread, cold cuts, and so on). Prepare his income statement for him.

3. Kyle Grant owns the Drug King Drugstore. In the first quarter, he prepared his income statement. His income from filling prescriptions was $18,354. His income from the rest of the store was $4,562. His rent for the building was $5,490, insurance was $857, utilities were $1,232, wages were $3,178, and inventory was $8,450. What does his income statement look like?

4. Jessica Fortuna has a small coffee shop in town. She wrote up an income statement for last month. She pays $1,250 rent on her shop. She paid $355 in insurance, $798 in utilities, $1,200 in wages, and $5,560 in inventory. Last month, she took in $10,576. What does her income statement look like?

5. Mrs. Helen Wait owns the Mosquito Bite Outdoor Store. She wrote up an income statement for last month. She pays $3,180 rent on her shop. She paid $675 in insurance, $1,363 in utilities, $5,600 for wages, and $8,450 for inventory. Last month the store took in $23,218. Prepare an income statement for her.

Unit 3: Inventory

Any business needs to know how many and what kind of goods it has on hand to be sold or used. Counting the number and types of goods on the shelves, however, may be difficult, especially if the business has multiple facilities and multiple types of goods in each facility. The number and type of goods at any one time is called the **inventory**. Businesses count their inventories yearly, semiannually, monthly, even weekly. Large chain stores are able to keep track of their inventory using computers and UPC codes on items for sale.

COST OF GOODS SOLD

Knowing what goods are on the shelves is not enough, however. A business owner must know how many goods have been sold, how much the sold items are worth, and how many goods remain on the shelves still available for sale. For instance, a vacuum cleaner store had bought 15 vacuum cleaners and has sold 8 of them. Each vacuum cleaner costs $275.00. The cost of the entire inventory is $4,125.00. Only 7 remain on the shelves. They are worth $1,925.00. The cost of goods sold is $4,125 − $1,925 = $2,200.

Of course, inventory is not as simple as that. Items can be stolen or broken. Stores have sales at various times during the year, and so goods on the shelves are sold at different prices. Additionally, goods are bought at different prices during the inventory period. Because of this, there are several ways of calculating the cost of goods sold at a store, at several stores, or nationwide (in the case of nationwide chains). We'll look at a few of these inventory methods.

SPECIFIC IDENTIFICATION INVENTORY METHOD

The first kind of inventory method is the **specific identification inventory method**. In this method, we find the total cost of the items purchased for the period and subtract the items remaining on the shelves at the end of the period.

 Great Pumpkin Patch (GPP) dolls are selling pretty well this year. Bela Van Damme, owner of The Big Doll Store, had to buy GPP dolls in an accelerated schedule, as follows:

Purchase Date	Dolls Bought	Cost Per Doll	Number of Dolls Left
Sept 10	18	$25	7
Sept 25	35	$23	13
Oct 20	42	$21	19
Nov 8	23	$25	0

What was the cost of goods sold during the fall?

SOLUTION

First find the cost of all the goods bought:

$$18 \times 25 = 450$$
$$35 \times 23 = 805$$
$$42 \times 21 = 882$$
$$23 \times 25 = 575$$

Now add all the totals:

$$450 + 805 + 882 + 575 = 2{,}712$$

Total cost of the GPP dolls is $2,712.00.

Find the cost of the dolls remaining:

$$7 \times 25 = 175$$
$$13 \times 23 = 299$$
$$19 \times 21 = 399$$
$$0 \times 25 = 0$$

Now add these totals:

$$175 + 299 + 399 + 0 = 873$$

Total cost of the remaining dolls, is $873.00.

Finally subtract the totals:

$$2{,}712 - 873 = 1{,}839$$

The cost of the goods sold for that period is $1,839.00.

WEIGHTED-AVERAGE METHOD

A second inventory method is the **weighted-average method**. In this method, the weighted average of the remaining items is determined. That number is then subtracted from the total cost.

Let's use the figures from the first example since we have already calculated the total cost. That figure was $2,712.00.

SOLUTION

Now divide this figure by the sum of the dolls bought:

$$\frac{2{,}712.00}{18 + 35 + 42 + 23} = \frac{2{,}712.00}{118} = 22.9830$$

The weighted average of the price of each GPP doll is $22.98. Now multiply this number by the number of dolls left:

$$22.98 \times (7 + 13 + 19 + 0) = 22.98 \times 39 = 896.22$$

The weighted value of the dolls still on the shelves is $896.22.

This number is subtracted from the total cost:

$$2{,}712 - 896.22 = 1{,}815.78$$

Thus the cost of the dolls sold during this period is $1,815.78.

RETAIL INVENTORY METHOD

Another method of tracking inventory is the **retail inventory method**. This method is very often used by department stores as it lends itself well to tracking a lot of different items. It is also relatively easy to calculate because the figures needed in the calculation are generally readily available to department store executives. In this method, the ratio of the cost (invoice price) of the goods to retail (selling) price is obtained. That fraction or decimal is multiplied by the actual retail sales made during the inventory period.

 Neisner's is a discount department store and wants to do an inventory of the goods in all its stores throughout the northeast. The cost (invoice price) of goods for the inventory period was $34,850.00, and the retail (selling) price was $48,790.00. The actual (retail) sales during that period were $42,573.00. What was the cost of goods during that period?

SOLUTION

First determine the ratio of the invoice price to the selling price:

$$\frac{34,850}{48,790} = 0.7142$$

Round this to 0.71.

Now multiply this decimal by the retail sales of that period:

$$42,573 \times 0.71 = 30,226.83$$

The cost of goods during that period was $30,226.83.

 E. J. Corvette is a department store and must do its an end-of-the-year inventory. The cost (invoice price) of goods for the year was $122,730.00, and the retail (selling) price was $241,611.00. The actual (retail) sales for the year were $195,864.00. What was the cost of goods that year?

SOLUTION

First determine the ratio of the invoice price to the selling price:

$$\frac{122,730}{241,611} = 0.5079$$

Round this to 0.51.

Now multiply this decimal by the retail sales of that period:

$$195,864 \times 0.51 = 99,890.64$$

The cost of goods during that period was $99,890.64.

INVENTORY TURNOVER RATE

The **inventory turnover rate** (ITR) helps store managers or executives to know when to order more merchandise. It is a simple concept. A turnover rate of 3 means that merchandise has been replaced (bought from the manufacturer *and* sold to the consumer) three times during the inventory period. Periods could be as short as a month or as long as a year. A low rate means that merchandise is not selling. The managers or executives must determine *why* it is not selling. The item may be priced too high, the item may break too easily, or consumers may just not like the item. Maybe the item is not being advertised enough or the ad campaign is not doing its job. A high turnover rate, on the other hand, may mean that the item is priced too low with respect to the competition. Profits for the company may be suffering as a result. The company may not be purchasing enough of the item to meet demand.

Although the ITR is a simple concept, there are no easy rules by which to guide the store manager. Generally, for perishable goods (produce, for instance), a high ITR is absolutely a must or the goods will all spoil. For nonperishable goods, many managers compare their ITR to that of other retailers of the same commodity, if they can get that rate from others.

To calculate the ITR, we use retail sales and both the beginning and the ending inventory prices for the period. Here are the formulas we use to calculate the ITR:

$$AIR = \frac{BIR - EIR}{2}$$

AIR — Average inventory retail
BIR — Beginning inventory retail
EIR — Ending inventory retail

Then we calculate the ITR:

$$ITR = \frac{\text{Net sales}}{AIR}$$

 EXAMPLE 5   The Monster Camper is a camping store and wants to calculate its inventory turnover rate for the first half of 2008. The (net) retail sales during that period was $12,531.00. The beginning inventory retail was $6,611.00. The ending inventory retail was $1,432.00. What was the ITR for the first half of 2008?

SOLUTION

First calculate the AIR:

$$AIR = \frac{6,611 + 1,432}{2} = \frac{8,043}{2} = 4,021.50$$

We round the AIR to 4,022.

Now divide the AIR into the retail sales to get the ITR:

$$ITR = \frac{12{,}531}{4{,}022} = 3.115$$

Round the ITR to 3. For the Monster Camper, the ITR is 3 for the first half of 2008.

EXERCISES

Using the figures provided, calculate inventory using the method indicated.

Purchase Date	Ship Models Bought	Cost of Ship Models	Number of Ship Models Left
Jan 10	20	$46	11
Feb 8	42	$40	10
Mar 14	30	$42	17
Apr 12	25	$46	0

Barnacle Bob's nautical curios shop bought ship models according the above table.

1. Find the cost of the goods available for sale.
2. Find the cost of goods sold using the specific identification method.
3. Find the cost of goods sold using the weighted-average method.

Purchase Date	Sand Cloaks Bought	Cost of Sand Cloaks	Number of Sand Cloaks Left
June 8	17	$40	6
June 21	20	$35	11
July 10	34	$33	16
July 30	20	$38	0

Greg LaBarge owns the Boardwalk Bums Beach Gear Hut. Sand Cloaks were very popular this year. He ordered them according to the above table.

4. Find the cost of the goods available for sale.
5. Find the cost of goods sold using the specific identification method.
6. Find the cost of goods sold using the weighted-average method.

Kent Webb has a contact lens business. During the third quarter, he looked at his books and saw that his cost (invoice price) of all his lenses and other items (cases, drops, and so on) was $45,785.00. The retail (selling) price of these goods was $68,677.50. His overall sales for the inventory period were $62,962.25.

7. Using the retail inventory method, find the cost of goods during that period.

At the Ham Hocks supermarket, Dave LaPorta, the manager, does a month-by-month inventory. During the month of February, he had all departments report. He learned that his cost (invoice price) of all his inventory was $64,564.00. The retail (selling) price of these goods was $83,933.20. His overall sales for the inventory period were $75,853.31.

8. Using the retail inventory method, find the cost of goods during that period.

John has a number of microprocessors in stock at his business to program and supply end users. He did an inventory for the spring season and found that the (net) retail sales for the inventory period was $82,435.00. The beginning inventory retail was $40,832.00. The ending inventory retail during that period was $32,680.00.

9. What was the inventory turnover rate during that period?

Harry sells baby clothes. He did an inventory for his busy season (Christmas). He calculated that the (net) retail sales for the inventory period was $42,525.00. The beginning inventory retail was $20,532.00. The ending inventory retail during that period was $17,595.00.

10. What was the inventory turnover rate during that period?

Unit 4: Depreciation

A landscape company buys lawnmowers for the summer months and snowplows for the winter months. A dentist buys a drill. Many businesses buy office furniture. These are assets of the business, but they wear out over time. Lawnmowers wear out. Snowplows break. The drill burns out its motor. The office furniture breaks down over time. Most useful items that are used by a company to carry on its business will wear out over time This is called **depreciation**. The useful life of the item is called the **estimated lifetime** of the asset. For tax purposes, the government allows the cost of these assets to be deducted from the profits of the business but not the whole cost of the asset in one year. The federal government will allow a business to deduct a portion of the cost of the asset each year over the **estimated lifetime** of the asset. Every year, the value of the item goes down by the depreciation amount. This value (cost – depreciation) is equal to the **book value**.

We will look at three ways to calculate the amount of depreciation over the **useful life** of the asset: the straight-line method, the sum-of-the-years-digits method, and the double-declining balance method.

STRAIGHT-LINE METHOD

The **straight-line method** is the one most often used because it is the simplest method for calculating depreciation. Under this method, the cost of the item (minus the scrap or resale value) is divided by the number of years of the useful life of the item. The scrap value is what the item would be worth to cut it up for scrap metal. The formula for calculating depreciation using the straight-line method is:

$$\text{Depreciation} = \frac{\text{Cost} - \text{Resale value}}{\text{Estimated lifetime}}$$

Pasquale has a landscape business. He buys a lawnmower for $15,000.00. After 5 years, the scrap value of the mower is $3,000.00. What is the depreciation amount per year?

SOLUTION

First subtract the resale value from the cost:

$$15,000 - 3,000 = 12,000$$

Now divide by the estimated lifetime of the mower:

$$x = \frac{12,000}{5} \Rightarrow x = 2,400$$

This means that every year, Pasquale can deduct $2,400.00 in depreciation off his taxes for the lawnmower.

An offset printing press is bought at a newspaper for $150,000.00. After 15 years, the scrap value of the printing press is $25,000. What is the depreciation amount per year?

SOLUTION

First subtract the resale value from the cost:

$$150,000 - 25,000 = 125,000$$

Now divide by the estimated lifetime of the mower:

$$x = \frac{125,000}{15} \Rightarrow x = 8,333.33$$

This means that every year, the newspaper can deduct $8,333.33 in depreciation off its taxes for the offset printing press.

SUM-OF-THE-YEARS-DIGITS METHOD

A slightly more involved way of calculating depreciation is the **sum-of-the-years-digits method**. Companies like to use this method because it shows, in a more realistic way, how equipment depreciates. Equipment becomes less and less efficient over time.

The estimated lifetime of the item is counted in years, for example 5. Then the years are numbered 1, 2, 3, 4, and 5. These become the numerators of five fractions, with the denominator being the sum of these digits: $1 + 2 + 3 + 4 + 5 = 15$. We now have 5 fractions:

$$\frac{1}{15} \quad \frac{2}{15} \quad \frac{3}{15} \quad \frac{4}{15} \quad \frac{5}{15}$$

To calculate the depreciation, multiply the cost of the machine (minus the scrap value) by the fractions in reverse order. That way, the largest fraction of the value of the machine is deducted the first year and so forth down to the end of its usable life. Let's use these fractions in the following example.

 EXAMPLE 3 Yani bought a van for his house-cleaning service. He paid $28,000.00 for the van. It has an estimated lifetime of 5 years and a scrap value of $3,000.00. Calculate the sum-of-the-years-digits depreciation for this van.

SOLUTION

Use fractions:

$$\frac{1}{15} \quad \frac{2}{15} \quad \frac{3}{15} \quad \frac{4}{15} \quad \frac{5}{15}$$

The cost minus the scrap value is:

$$28,000 - 3,000 = 25,000$$

Multiply, in reverse order, the fractions by the 25,000. The book value is the value minus the year's depreciation:

Year	Fraction	Book Value
Year 1:	$\frac{5}{15} \times 25,000 = 8,333.33$	$16,666.67
Year 2:	$\frac{4}{15} \times 25,000 = 6,666.67$	$10,000.00
Year 3:	$\frac{3}{15} \times 25,000 = 5,000.00$	$5,000.00
Year 4:	$\frac{2}{15} \times 25,000 = 3,333.33$	$1,666.67
Year 5:	$\frac{1}{15} \times 25,000 = 1,666.67$	$0.00

The sum of the book values is 25,000. So the full value is accounted for (minus the scrap value).

In each year, Yani will deduct the amount determined in the table. He will get a larger deduction in the first year than in the last year.

DOUBLE-DECLINING BALANCE METHOD

This method is similar to the sum-of-the-digits method in that it assumes the greatest amount of depreciation happens in the earlier years. The item depreciates at a lesser rate over time. However, in the double-declining balance method, the item doesn't depreciate below its scrap (or resale) value. First, the number one (100%) is divided by the estimated lifetime of the item. The result is always decimal. Then this number is doubled. Here is the formula:

$$\frac{100\%}{\text{Estimated lifetime}} \times 2 = \text{Depreciation rate}$$

This decimal is then multiplied by the original cost of the item. The result becomes the first year's depreciation value. That depreciation value is subtracted from the original cost, and that becomes the book value. Unlike with the sum-of-the-digits method, we don't subtract the scrap value. As with the sum-of-the-digits method, it's easiest to see this with an example.

 José manages a restaurant. He bought new furniture that cost $1,500.00 and has a useful life of 4 years. Its scrap value is $70. Calculate the depreciated values over the 4 years using the double-declining balance method.

SOLUTION

First calculate the double-declining balance rate:

$$\frac{1.00}{4} \times 2 = 0.5$$

Use 0.5 to calculate the depreciation each year. The book value is the value minus the year's depreciation:

Year	Fraction	Book Value
Year 1:	0.5 × 1,500 = 750	1,500 − 750 = $750.00
Year 2:	0.5 × 750 = 375	750 − 375 = $375.00
Year 3:	0.5 × 375 = 187.5	375 − 187.50 = $187.50
Year 4:	0.5 × 187.5 = 93.75	187.50 − 93.75 = $93.75

The final value of the furniture after 4 years is $93.75.

EXERCISES

1. Carol has a motorcycle repair business. She buys a frame-straightening machine for $55,000.00. After 8 years, the scrap value of the frame-straightening machine is $12,000.00. Using the straight-line method, calculate the depreciation amount per year.

2. Bill King has a boat drydock business. He buys a floating drydock that floats under the boat and then lifts up the boat for the winter months. He pays $120,000.00 for the drydock. After 16 years, the scrap value of the drydock is $35,000.00. Using the straight-line method, calculate the depreciation amount per year for Mr. King's drydock.

3. Jacqueline owns the Bagel Bakery. She buys an espresso machine for $550.00. After 10 years, the resale value of the espresso machine is $60.00. Using the straight-line method, calculate the depreciation amount per year.

4. Tom Williams owns a production (event-staging) business. He buys a stage for $65,000.00. After 7 years, the resale value of this stage is $7,000.00. Using the straight-line method, calculate the depreciation amount per year.

5. Adore got a deal on an oven for her pie-baking business. She paid $29,500.00 for it. It has an estimated lifetime of 9 years and a resale value of $11,500.00. Calculate the sum-of-the-years-digits depreciation for this oven.

6. Jini dispatches ambulances throughout the city as part of First Reaction Medical Transport. The city paid $22,000.00 for the phone-switching machine she uses. It has an estimated lifetime of 12 years and a scrap value of $4,500.00. Calculate the sum-of-the-years-digits depreciation for this switching machine.

7. Walter has a log-splitting business. He paid $13,500.00 for a pneumatic log splitter. It has an estimated lifetime of 9 years and a scrap value of $4,000.00. Calculate the sum-of-the-years-digits depreciation for this oven.

8. Miguel has a body shop for automobiles. He bought a spray-painting machine for $10,500.00, and it has a useful life of 10 years. Its scrap value is $1,100. Calculate the depreciated values over the 10 years using the double-declining balance method.

9. Tom Gallagher owns a tugboat business. He bought a new tugboat that cost $105,000.00 and has a useful life of 8 years. Its scrap value is $10,500. Calculate the depreciated values over its useful life using the double-declining balance method.

10. Greg has a restaurant in town. He bought a walk-in freezer for $18,500.00 that has a useful life of 9 years. Its scrap value is $1,950. Calculate the depreciated values over its estimated lifetime using the double-declining balance method.

Unit 5: Review of Chapter 11

Questions 1 and 2 are balance sheet exercises. Calculate the total assets, liabilities, and net worth of the following companies.

1. Balance Sheet:

		We-Dig-U Balance sheet		
Assets:				
	Cash on hand:		$ 6,753	
	Accounts receivable:		$ 11,943	
	Vehicles:		$ 87,000	
	Furniture:		$ 5,632	
	Inventory:		$ 2,500	
	Office supplies:		$ 435	
	Total:		$	
Liabilities:				
	Rent:		$ 2,500	
	Vehicle debt:		$ 19,500	
	Payroll:		$ 48,765	
	Accounts payable:		$ 8,665	
	Overhead:		$ 7,870	
	Total:		$	
	Equity (net worth):			

2. Balance Sheet:

		Hi-Wire Electronics Balance sheet			
Assets:					
	Cash on hand:		$ 5,653		
	Accounts receivable:		$ 7,629		
	Building:		$ 125,000		
	Vehicles:		$ 22,500		
	Furniture:		$ 8,831		
	Inventory:		$ 51,565		
	Office supplies:		$ 4,835		
	Total:		$		
Liabilities:					
	Mortgage:		$ 87,450		
	Vehicle debt:		$ 11,250		
	Payroll:		$ 42,560		
	Accounts payable:		$ 3,547		
	Overhead:		$ 6,534		
	Total:		$		
	Equity (net worth):				

Questions 3 and 4 are income statement exercises. Calculate the total revenues, the total liabilities, and the net profit of each company.

3. Kathy Kim manages an apartment building in Manville. For the fourth quarter, she wrote an income statement. She had rental income of $22,450. She paid $6,550 mortgage on the building, $1,030 in insurance, $1,575 in utilities, and $1,235 in building repairs. What does her income statement look like?

	Kathy Kim's Income Statement			
Revenues:				
	Rental income:		$	
	Total revenues:		$	
Expenses:				
	Mortgage:		$	
	Insurance:		$	
	Utilities:		$	
	Building repairs:		$	
	Total expenses:		$	
	Net profit =			

4. Mr. Lonni Hanks owns a corner convenience store. In June, he prepared an income statement. He had gross sales of $12,655. He paid $1,200 for rent of his store, $135 in insurance, $432 in utilities, and $752 in incidentals (advertising, mostly). What does his income statement look like?

	Mr. Lonni Hanks's Income Statement			
Revenues:				
	Gross sales:		$	
	Total revenues:		$	
Expenses:				
	Rent:		$	
	Insurance:		$	
	Utilities:		$	
	Incidentals:		$	
	Total expenses:		$	
	Net profit =			

Questions 5–13 deal with inventory methods.

Branford Dameon, owner of The Cheese Shop, bought marble cutting boards to sell with his cheese as follows:

Purchase Date	Cutting Boards Bought	Cost of Cutting Board	Number of Cutting Boards left
Sept 15	25	$12	9
Sept 30	30	$10	12
Oct 15	35	$ 9	20
Oct 30	30	$10	0

5. What is the cost of the goods available for sale?

6. What is the cost of goods sold using the specific identification method?

7. What is the cost of goods sold using the weighted-average method?

Marilyn Lee manages The Corker Wine Shop. An exceptionally fine pinot grigio from Italy was available, and she ordered several cases according to the following table:

Purchase Date	Cases of Wine Bought	Cost of Cases	Number of Cases Left
March 9	12	$120	5
April 1	21	$108	15
April 29	30	$ 96	18
May 17	25	$108	0

8. What is the cost of the goods available for sale?

9. What is the cost of goods sold using the specific identification method?

10. What is the cost of goods sold using the weighted-average method?

11. Jean LaPlace has a ladies accessory shop, The Well-Stuffed Purse. During the first quarter, she looked at the books and calculated that the cost (invoice price) of all the goods was $68,531.00. The retail (selling) price of these goods was $122,579.00. The overall sales for the first quarter were $97,982.47.

    Using the retail inventory method, find the cost of goods during that period.

12. Ralph Oberg, the manager of A Long Walk shoestore, does a bimonthly inventory. During September and October, he checked his books and learned that the cost (invoice price) of all his inventory was $57,723.50. The retail (selling) price of these goods was $100,035.70. His overall sales for the inventory period were $83,382.58.

    Using the retail inventory method, find the cost of goods during that period.

13. Lori Orginski has a coffee shop. She did an inventory for the end of the year (October to December). She found that the (net) retail sales for that period was $91,375.40. The beginning inventory retail was $33,375.20. The ending inventory retail during that period was $17,391.60.

    What was the inventory turnover rate during that period?

Questions 14–22 deal with depreciation methods.

14. Rob Heirich has a plumbing business. He works with sheet metal as well. He buys a sheet metal–shaping machine for $85,000.00. After 20 years, the scrap value of the sheet metal–shaping machine is $12,000.00.

    Using the straight-line method, calculate the depreciation amount per year for Mr. Heirich's sheet metal–shaping machine.

15. Charlie Crane is general manager of a small pharmaceutical business. He purchases a pill-sorting and -packaging machine for $125,000.00. After 18 years, the resale value of the pill-sorting machine is $21,000.00.

    Using the straight-line method, calculate the depreciation amount per year of the pill-sorting and packaging machine.

16. Barbara McDermott owns I Spy, an investigative business. She buys a van for $45,000.00 for her gear. After 6 years, the resale value of this van is $4,000.00.

    Using the straight-line method, calculate the depreciation amount per year of the van.

17. Joyce Orbach bought a rug-cutting machine for $32,560.00 for her rug business. It has an estimated lifetime of 7 years and a resale value of $3,100.00.

    Calculate the sum-of-the-years-digits depreciation for this rug-cutting machine.

18. James Van der Peen works for the town as a firefighter. The city paid $92,000.00 for the hook and ladder truck he works on. It has an estimated lifetime of 6 years and a scrap value of $24,500.00.

    Calculate the sum-of-the-years-digits depreciation for this truck.

19. Stanley Wojciechowski is a lab technician. The lab he works in paid $27,565.00 for an X-ray machine. It has an estimated lifetime of 9 years and a scrap value of $8,500.00.

    Calculate the sum-of-the-years-digits depreciation for this X-ray machine.

20. Maria DeJesus is an opthamologist (eye doctor). She bought an eye-examining machine for $26,700.00, and it has a useful life of 15 years. Its scrap value is $3,300.

    Calculate the depreciated values over the 10 years using the double-declining balance method.

21. Lori Feldman owns an ice cream store. She bought a new ice cream–freezing machine that cost $31,250.00 and has a useful life of 12 years. Its scrap value is $3,340.00.

    Calculate the depreciated values over the useful life of the ice cream–freezing machine using the double-declining balance method.

Answers to Exercises

Chapter 1, Unit 1

Exercise A **1.** $201.19 **2.** $1,595.53 **3.** $1,178.60 **4.** $3,119.20
5. $9,305.80 **6.** $5,588.77 **7.** $17,099.30 **8.** $40,199.27
9. $66,394.53 **10.** $57,409.36

Exercise B **11.** $391.24 **12.** $406.55 **13.** Sales: $78,442.30;
Commission: $7,844.24 **14.** 546 **15.** 17,908
16. Subtotal: $93.50; Total: $100.98

Exercise C **17.** $5,211.60 **18.** $1,622.93 **19.** $80.50 **20.** $364.68
21. $312.09

Exercise D **22.** 35 **23.** 225 **24.** 323 **25.** 21,962 **26.** 88,065
27. 130,935 **28.** 175,968 **29.** 91,277 **30.** 150,720

Exercise E **31.** Adams: 50; Adman: 45; Burke: 41; Curtis: 44; Dellman: 44;
Evans: 46
Mon.: 53; Tues.: 49; Wed.: 61; Thurs.: 52; Fri.: 55;
Grand Total: 270
32. 01: $70.87; 02: $83.70; 03: $84.93; 04: $77.01;
05: $105.84; 06: $102.90; 07: $88.23; 08: $111.78;
09: $77.51; 10: $106.85 Fed. Tax: $462.26;
FICA Tax: $158.35; State Tax: $106.15; City Tax:
$67.94; Pension: $83.75; Health Plan: $31.17;
Grand Total: $909.62
33. Grocery: $13,556.32; Produce: $6,248.82; Dairy: $10,080.39;
Meat: $8,688.70; Deli: $5,339.13; Nonfood: $4,625.11
Mon.: $7,631.68; Tues.: $7,477.31; Wed.: $6,905.38;
Thurs.: $7,162.70; Fri.: $6,848.84; Sat.: $6,371.98;
Sun.: $6,140.58; Grand Total: $48,538.47

34. Ladies' Wear: $9,220.01;   Men's Wear: $7,088.14;   Children's
Wear: $3,838.75;   Appliances: $10,960.37;   Furniture:
$13,891.69;   Toys: $3,819.78   Cash: $15,840.78;
Charge: $16,595.11;   C.O.D.: $16,382.85;   Grand Total: $48,818.74

Exercise F   **35.** $2,785.09   **36.** $4,501.02   **37.** $2,445.37   **38.** $547.93
39. $220.34   **40.** $199.52   **41.** $195.52

Chapter 1, Unit 2

Exercise A   **1.** 42   **2.** 43   **3.** 20   **4.** 49   **5.** 57   **6.** 79
7. 89   **8.** $228.35   **9.** $67.19   **10.** $278.07   **11.** $78.79
12. 983   **13.** 4,924   **14.** $1,890.88   **15.** $19,217.08
16. $8,918.93   **17.** 18,717   **18.** 7,926   **19.** 9,086
20. 39,065   **21.** 958   **22.** 235   **23.** 68   **24.** 1,266
25. 42,957   **26.** 17,838

Exercise B   **27.** $172.98   **28.** $194.95   **29.** $219.07   **30.** $216.07
31. $334.13   **32.** $280.75   **33.** $201.28   **34.** $316.25
35. $1,964.89   **36.** $808.84   **37.** $953.71   **38.** $2,094.52
39. $1,790.70   **40.** $215.85   **41.** $207.92

Exercise C   **42.** 247   **43.** $1,853   **44.** $118.88   **45.** $194
46. $30,292.81

Exercise D   **47.** $24.95   **48.** $30.87   **49.** $87.30   **50.** 14,165
51. $3.94   **52.** $4.38   **53.** 3,085   **54.** 27,892
55. $463.60   **56.** .982   **57.** 12,105,081   **58.** $3,793.76

Exercise E   **59.** Suits: 440;   Slacks: 207;   Sports Jackets: 292;   Coats: 263;
Sweaters: 269;   Sport Shirts: 283
Stock: 2,686;   Sold: 932;   Grand Total: 1,754
60. Appliances: $4,763.80;   Furniture: $18,530.14;   Lamps: $3,213.79
Sales: $29,376.18;   Returns: $2,868.45;   Grand Total: $26,507.73
61. A: $5,589.85;   B: $1,369.25;   C: $871.86;   D: $2,772.22;
E: $3,203.95;   F: $3,857.55;   G: $2,106.90
Original Price: $30,515.00;   Amount of Reduction: $10,743.42;
Grand Total: $19,771.58
62. Alan: $168.52;   Ambrose: $196.39;   Baker: $164.47;
Buntel: $158.62;   Campbell: $183.79;   Chisholm: $169.55;
Fulton: $167.72   Gross Pay: $1,466.28;
Deductions: $257.22;   Grand Total: $1,209.06
63. A: $207.23;   B: $282.85;   C: $260.03;   D: $214.18;
E: $270.52;   F: $241.65;   G: $244.43;   H: $283.53
Gross Pay: $2,681.51;   Deductions: $677.09;
Grand Total: $2,004.42

64. Grocery: $2,183.95;   Vegetables: $1,715.47;   Dairy: $1,180.80;
Delicatessen: $1,111.16;   Nonfood: $893.29;   Meat: $1,887.93
Gross Profit: $13,953.21;   Overhead: $4,980.61;
Grand Total: $8,972.60

Chapter 1, Unit 3

Exercise A   **1.** 12,964   **2.** 19,278   **3.** 1,695,421   **4.** 8,154.12   **5.** 254.8
6. 163.552   **7.** 79,130   **8.** 2,656.5625   **9.** 639.5175
10. 100.4523   **11.** 169.158   **12.** 10.515875   **13.** 1,886.34375

Exercise B   **14.** #1302: $330.30;   #1567: $1,107.84;   #158: $1,290.60;
#758: $229.14;   Total: $2,957.88
15. #1728: $474.00;   #2071: $621.00;   #509: $448.20;
#1249: $137.90;   #1415: $334.80;   Total: $2,015.90
16. #1106: $228.20;   #1108: $937.50;   Tiles, Yellow:
$1,433.44;   Tiles, Blue: $960.75;   #708: $1,072.50;
#710: $731.25;   Total: $5,363.64
17. #307: $1,754.55;   #461: $7,792.05;   #107: $3,043.25;
#423: $7,175.00;   #613: $6,526.05;   #707: $8,788.50;
#321: $723.20;   Total: $35,802.60
18. White: $1,923.25;   Pink: $1,723.75;   Blue: $2,912.50;
Green: $1,487.50;   Sage: $1,633.25;   Orange: $2,583.75;
Total: $12,264

Exercise C   **19.** $3,194

20. $613.80   **21.** $2,227.50   **22.** $282.50   **23.** $21,506.25

Exercise D   **24.** $84.86   **25.** $10   **26.** $21.96   **27.** $23   **28.** $110.90
29. 3,000   **30.** $15.25   **31.** $229.31   **32.** $42.90
33. $8,976.79

Exercise E   **34.** Alcohol, Wood: $291.94;   Thinner: $422.63;
Solvent #5: $721.21,   Solvent #8: $710.72;
Alcohol, Denatured: $777.95;
Lacquer: $817.88;   Total: $3,742.33
35. Potassium: $97.30;   Magnesium: $108.86;
Sulfur: $169.20;   Total: $375.36
36. Wheat: $1,624.35;   Rice: $1,369.77;
Soya Bean: $1,349.93;   Corn: $1,161.53;
Oat: $1,089.92;   Total: $6,595.50

37.

$1	$100	$1,000	$10,000	$100,000
$ 356,248	$ 356,200	$ 356,000	$ 360,000	$ 400,000
474,563	474,600	475,000	470,000	500,000
1,263,438	1,263,400	1,263,000	1,260,000	1,300,000
835,726	835,700	836,000	840,000	800,000
452,375	452,400	452,000	450,000	500,000
168,935	168,900	169,000	170,000	200,000
244,569	244,600	245,000	240,000	200,000

Exercise F   **38.** $272.01   **39.** $411.45

Exercise G   **40.** 4.27   **41.** 53.4   **42.** $5,150   **43.** $2,800   **44.** $9.00
45. $2.50   **46.** $9,045   **47.** $3,150   **48.** 90   **49.** 8,700
50. 4,500   **51.** 2,700

Exercise H   **52.** Screwdrivers: $750;   #19 Nails: $50;   #22 Nails: $530;
#15 Nails: $3,900;   Wire: $2,482;   Total: $7,712
53. Sweaters: $955;   Blouses: $2,475;   Short Jackets: $850;
Full-length Jackets: $960;   Total: $5,240
54. Pencils: $300;   Erasers: $240;   Pens: $3,600;
12" Rulers: $1,060;   8" Rulers: $1,500;   Total: $6,700

Chapter 1, Unit 4

Exercise A   **1.** 1186   **2.** 0.827   **3.** 15.31238   **4.** 7   **5.** 624
6. 1,539.0935   **7.** 232.1   **8.** 216.01639   **9.** 1.5225071
10. 1.721   **11.** 0.2130963   **12.** 3,614,899.3   **13.** 314,132.35
14. .00001

Exercise B   **15.** Pencils: $.06;   Pads: $1.31;   Erasers: $.08;   Ribbons: $.65;
#10 Envelopes: $9.17;   #6 Envelopes: $4.55
16. Blue: 126;   Plum: 132;   Gold: 132;   Red: 116;   Royal Blue: 125
17. Carpeting: $6;   Tiles: $5;   Fabric: $5;   Wallpaper: $7;   Paint: $5
18. $103.15; $182.53; $104.90; $237.86; $201.58; $135.01;
$159.12; $149.12; $157.94; $110.48
19. 1: $1.22;   2: $3.32;   3: $3.57;   4: $2.88;   5: $3.05;
6: $5.45;   7: $4.48;   8: $3.24;   9: $3.22;   10: $6.31
20. 1: $4.09;   2: $5.04;   3: $3.57;   4: $2.90;   5: $3.03;   6: $13.91;
7: $10.54;   8: $15.85;   9: $13.89;   10: $15.07

Exercise C   **21.** $11   **22.** $5   **23.** $7.25   **24.** $23

Exercise D   **25.** $2.35   **26.** $.06   **27.** 4.21   **28.** $.02   **29.** 82.67
30. .063   **31.** .35   **32.** $3.24   **33.** $.12   **34.** $5   **35.** $1,800
36. $13,215   **37.** $51,000

Exercise E　**38.** Pens: $45.13;　Notebooks: $90.30;　Erasers: $29.38;
　Pencils: $50.10;　Pens: $60.63
　39. Rice: $1,577.52;　Wheat: $1,503.53;　Barley: $1,358.73;
　Peanuts: $677.56;　Walnuts: $2,159.41
Exercise F　**40.** $47.50　**41.** $18.56

Chapter 1, Unit 5

Exercise A　**1.** $\frac{1}{3}$　**2.** $\frac{3}{10}$　**3.** $\frac{6}{11}$　**4.** $\frac{3}{4}$　**5.** $\frac{3}{7}$　**6.** $\frac{20}{25}$　**7.** $\frac{25}{30}$

8. $\frac{44}{48}$　**9.** $\frac{28}{60}$　**10.** $\frac{33}{72}$　**11.** $3\frac{4}{5}$　**12.** $3\frac{7}{12}$　**13.** $3\frac{3}{8}$

14. $1\frac{7}{9}$　**15.** $8\frac{1}{2}$　**16.** $5\frac{5}{8}$　**17.** $1\frac{1}{13}$　**18.** $1\frac{1}{3}$　**19.** $\frac{31}{7}$

20. $\frac{37}{3}$　**21.** $\frac{47}{3}$　**22.** $\frac{27}{4}$　**23.** $\frac{19}{7}$　**24.** $\frac{47}{6}$　**25.** $\frac{38}{3}$

Exercise B　**26.** 1　**27.** $\frac{15}{20}$　**28.** 60¢　**29.** $\frac{1}{3}$

30. (a) $\frac{1}{3}$　(b) $10,000

31. $\frac{2}{5}$　**32.** (a) $\frac{3}{22}$　(b) $\frac{19}{22}$

33. $\frac{3}{4}$　**34.** $\frac{1}{3}$

Chapter 1, Unit 6

Exercise A　**1.** $1\frac{13}{15}$　**2.** $1\frac{5}{12}$　**3.** $1\frac{3}{4}$　**4.** $37\frac{151}{168}$　**5.** $70\frac{17}{60}$

6. $65\frac{7}{40}$　**7.** $59\frac{3}{10}$　**8.** $52\frac{29}{30}$　**9.** $\frac{1}{20}$　**10.** $\frac{11}{60}$

11. $\frac{1}{12}$　**12.** $15\frac{1}{8}$　**13.** $16\frac{11}{30}$　**14.** $4\frac{17}{24}$　**15.** $8\frac{1}{4}$

Exercise B　**16.** Alvarez: $46\frac{1}{4}$;　Baines: 46;　Belmore: 50;　Caldwell: $48\frac{3}{4}$;
　Carter: $48\frac{3}{4}$;　Cortez: $49\frac{3}{4}$;　Elton: $48\frac{3}{4}$

17. Alvarez: $48\frac{3}{4}$;　Baines: $49\frac{1}{2}$;　Belmore: $48\frac{3}{4}$;　Caldwell: 49;
　Carter: $50\frac{3}{4}$;　Cortez: $48\frac{3}{4}$;　Elton: $49\frac{3}{4}$

Exercise C　**18.** $20\frac{1}{4}$　**19.** $3\frac{5}{6}$　**20.** $17\frac{13}{48}$　**21.** $\frac{1}{6}$　**22.** $17\frac{5}{6}$

Chapter 1, Unit 7

Exercise A   **1.** $\frac{7}{12}$ **2.** $\frac{7}{54}$ **3.** $\frac{7}{15}$ **4.** $21\frac{2}{3}$ **5.** $13\frac{1}{5}$ **6.** $\frac{5}{6}$ **7.** $\frac{8}{9}$

8. $2\frac{2}{5}$ **9.** $1\frac{64}{77}$ **10.** $2\frac{1}{8}$ **11.** $\frac{19}{22}$ **12.** $10\frac{2}{3}$ **13.** 4

Exercise B   **14.** $3\frac{3}{4}$ **15.** 588   **16.** 20   **17.** $212.15

18. (a) $83.33   (b) $50   (c) $116.67

Chapter 1, Unit 8

1. $1,019.85   **2.** $1,632.93

3. Cash Sales: $137,950.02;   Charge Sales: $149,928.06

4. Sales: $116,850.54; Commissions: $8,179.53

5. $206.67   **6.** $6,050.52

7. A: $56.85;   B: $167.55;   C: $178.05;   D: $74.74;   E: $120.17;
F: $201.25;   G: $190.95
Original Price: $3,794.95;   Sales Price: $2,805.39;   Grand Total: $989.56

8. A: $149.00;   B: $143.01;   C: $46.00;   D: $206.00;   E: $290.00;
F: $62.98;   G: $96.05
List Price: $1,306.68;   Discount: $313.64;   Grand Total: $993.04

9. $274.24   **10.** $6,851.99

11. 4137: $1,255.91;   4332: $3,449.04;   4253: $12,610.07;
4315: $13,341.97;   4528: $14,882.88;   Amount: $46,514.27;
Cash Discount: $974.40;   Grand Total: $45,539.87

	Total Deductions	Net Pay		Deductions	Net Pay
12. 101	$190.25	$275.60	104	$167.49	$230.93
102	177.60	255.10	105	182.82	263.54
103	200.06	278.42	106	175.09	305.61

13. $14,736.15   **14.** $228.30

15. Rice: $2,256.07;   Corn: $2,276.99;   Lentil: $1,980.47;
Soya Bean: $2,072.09;   Sesame: $1,071.45;   Total: $9,657.07

16. Clips: $100;   Ribbons: $2,500;   Fluid: $1,400;   Envelopes: $360;
Pads: $1,460;   Total: $5,820

17. $110   **18.** $1,250   **19.** $7,780   **20.** $10,423.40   **21.** $1,897.20

22. $289.75   **23.** $44.71   **24.** 1,562   **25.** 46   **26.** 2.15

27. 4,213,000   **28.** $24.61   **29.** $22.57   **30.** $135.39   **31.** 505

32. 4,003   **33.** 20.004   **34.** 630.33   **35.** $.15   **36.** 6   **37.** 20

38. 2.467   **39.** 2   **40.** $65.91; $76.97; $108.02; $136.91; $206.54

41. Sugar: $28.01;   Coffee: $82.24;   Wheat: $1,130.47;   Peanuts: $1,996.44

42. $345.95   **43.** $219.62   **44.** $4,500

45. $\frac{1}{8}$ **46.** $\frac{4}{9}$ **47.** $\frac{1}{10}$ **48.** $\frac{36}{48}$ **49.** $\frac{36}{45}$ **50.** $2\frac{1}{4}$

51. $1\frac{67}{90}$ **52.** $80\frac{7}{8}$ **53.** $\frac{1}{15}$ **54.** $\frac{7}{12}$ **55.** $5\frac{1}{8}$ **56.** $\frac{3}{4}$

57. 176   **58.** 8,319   **59.** $\frac{14}{15}$ **60.** $\frac{22}{43}$ **61.** $\frac{69}{82}$

Chapter 2, Unit 1

Exercise A   **1.** 0.7   **2.** 0.5   **3.** 0.7   **4.** 0.88   **5.** 0.42   **6.** 0.72
7. 0.529   **8.** 0.278   **9.** 0.721

Exercise B   **10.** 0.656   **11.** 0.047   **12.** 0.8   **13.** 0.667   **14.** 0.20 or 0.2

Chapter 2, Unit 2

Exercise A   **1.** .29   **2.** .375   **3.** 1.00   **4.** .0798   **5.** .0001
6. .0675   **7.** 25%   **8.** 6%   **9.** 8,750%   **10.** 200%
11. 215%   **12.** 20%   **13.** .0825   **14.** .125
15. $254.80   **16.** $204.75   **17.** $180   **18.** $42.56
19. $13.90   **20.** $7.53

Exercise B   **21.** Subtotal: $1,167.00;   Sales Tax: $93.36;   Total: $1,260.36

Exercise C   **22.** 40¢   **23.** $33\frac{1}{3}\%$ **24.** 7.50%   **25.** $39,192
26. $763.75   **27.** $379.50

Exercise D   **28.** 87.5%   **29.** 19.2%   **30.** 25%   **31.** 55.6%   **32.** 20%
33. 37.5%   **34.** 16.7%   **35.** 147%   **36.** 50%   **37.** 58.4%
38. 39.4%   **39.** 5.3%   **40.** 57.4%   **41.** 23.1%   **42.** 20.6%
43. 153.9%   **44.** 21.1%   **45.** 21.2%   **46.** 42.7%
47. 176.6%   **48.** 30.9%   **49.** 34.3%   **50.** $205.71
51. $877.78   **52.** $1,107   **53.** 587.5   **54.** 9,840
55. $225.45   **56.** 448   **57.** $2,833.73   **58.** $72.86
59. 2,425.93

Exercise E   **60.** Sale Price: $153.17;   Reduction Percent: 35%
Reduction Amount: $132.13;   Reduction Percent: 40%
Sale Price: $114.30;   Reduction Amount: $61.55
Reduction Amount: $259.18;   Reduction Percent: 45%
Sale Price: $285.25;   Reduction Percent: 33%
Reduction Amount: $182.52;   Reduction Percent: 48%
Reduction Amount: $104.63;   Reduction Percent: 35%
Sale Price: $436.21;   Reduction Amount: $169.64

61. Amount of Sale: $80.92;   Sales Tax Amount: $4.86
Sales Tax Rate: 8%;   Sales Tax Amount: $9.53
Amount of Sale: $233.36;   Sales Tax Amount: $12.83
Amount of Sale: $178.61;   Sales Tax Amount: $14.74
Sales Tax Rate: 5%;   Amount of Sale: $206.17
Amount of Sale: $329.14;   Sales Tax Amount: $13.17
Sales Tax Rate: 4.5%;   Sales Tax Amount: $11.16
Amount of Sale: $321.31;   Sales Tax Amount: $20.89

Exercise F   **62.** 17.9%   **63.** 21.2%   **64.** 19.8%   **65.** 22.6%   **66.** 48.8%

Exercise G   **67.** $193.08   **68.** $356,666.67   **69.** $13,260

Chapter 2, Unit 3

Exercise A   **1.** $\frac{3}{4}$ **2.** $\frac{3}{1}$ **3.** $\frac{8}{1}$ **4.** $\frac{6}{25}$ **5.** $\frac{9}{1}$ **6.** $\frac{1.20}{1}$

7. $\frac{.1}{1}$ **8.** $\frac{.58}{1}$

Exercise B   **9.** $40   **10.** 108

Exercise C   **11.** 5   **12.** 30   **13.** 5   **14.** 32   **15.** .17   **16.** 315
17. 3,552   **18.** $35.25   **19.** $520

Exercise D   **20.** $3,275   **21.** $250

Chapter 2, Unit 4

1. .8   **2.** .8   **3.** .33   **4.** .60   **5.** .438   **6.** .639   **7.** $125

		Amount of Discount	Net Price
8.	A	$ 50.80	$ 76.20
	B	73.35	89.65
	C	149.60	122.40
	D	85.75	159.25
	E	122.98	163.02
	F	177.84	164.16
	G	318.66	359.34

	Amount of Interest	New Balance
9.	$108.41	$1,347.06
	153.90	1,732.32
	193.86	2,609.56
	248.18	2,819.99
	369.45	4,052.00
	410.22	3,930.82

10. $280   **11.** $413.17   **12.** (a) $10,909.25   (b) 20%
13. $11,207.69   **14.** $5,566.84

Chapter 3, Unit 1

Exercise A   **1.** 32   **2.** 320.8   **3.** 253.2   **4.** 458   **5.** 20.1

	Median	Mode		Median	Mode
6.	70	70	**9.**	47.5	40
7.	25	32	**10.**	180	170
8.	170	191			

Exercise B   **11.** $302.89   **12.** (a) $10,466.67   (b) $10,175   **13.** $195

Chapter 3, Unit 2

Exercise A   **1.** Sales for Appliance Department   **2.** $5,000
3. (a) August   (b) $37,000   **4.** May and September
5. February, March, May, July, August, and November
6. (a) July   (b) $35,000   **7.** October   **8.** July

Exercise B   **9.**

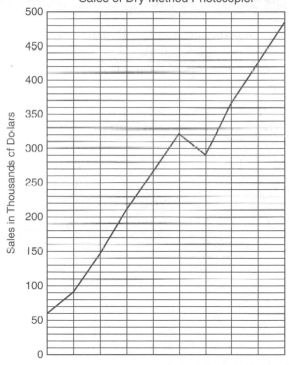

Photo-Tech Manufacturing Company
Sales of Dry-Method Photocopier

Chapter 3, Unit 3

1. 4 | 5, 8
  5 | 0, 2, 3, 5, 5, 8
  6 | 5, 7
    Median: 54   Mode: 55

2. 1 | 2, 3, 5, 5, 8, 8
  2 | 1, 3, 5, 6
    Median: 18   Modes: 15, 18

3. 1 | 5
  2 | 1, 2, 2, 8
  3 | 0, 7, 7, 7
  4 | 4, 5, 5
    Median: 33.5   Mode: 37

4. 5 | 3, 5, 9
  6 | 6
  7 | 3, 4, 5, 8
  8 | 3, 3, 4, 5, 5
  9 | 4, 6
    Median: 78   Modes: 83, 85

5. 6 | 9, 9
  7 | 5, 7
  8 | 1, 5, 9, 9, 9
  9 | 1, 5, 5
    Median: 87   Mode: 89

Chapter 3, Unit 4

Exercise A **1.** Thrifty Supermarket Sales by Department   **2.** $10,000
  3. (a) Grocery   (b) $88,000
  4. $42,000   **5.** Nonfood and Deli   **6.** Deli and Grocery
  7. 22.9%
  8. Grocery, Meat, Dairy, Produce, Nonfood, Deli
    (a) Deli   (b) Grocery and Dairy

Exercise B **9.** 2002   **10.** 450   **11.** 350   **12.** 36%

Exercise C **13.**

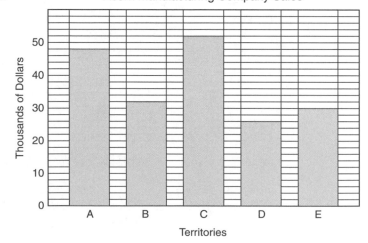

14.

Tuition Costs for 1 Year at a State College

Chapter 3, Unit 5

Exercise A **1.** Distribution of the Fernandez Income of $28,500   **2.** 100%
3. Charity   **4.** 59%   **5.** $6,270   **6.** $570

Exercise B **7.**   Bee Gee Appliance Company Sales

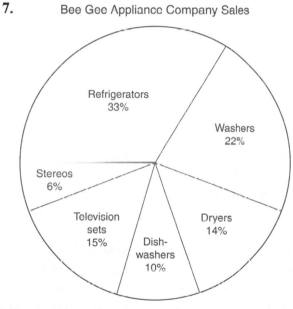

Chapter 3, Unit 6

1. 38   **2.** 347.2
3. Median: 75;   Mode: 70
4. (a) 1998   (b) $78,000
5. 1995

   6. $36,000; $42,000; $38,000; $44,000; $54,000; $70,000;
       $58,000; $78,000; $64,000; $58,000;
       Mean = $54,200; Median = $56,000; Mode = $58,000

Note that, with three-digit numbers, it is standard practice to put the hundreds and the tens digits to the left of the line.

   7. 42 | 3, 5, 9
       45 | 4, 5, 6
       55 | 2, 3, 6
       56 | 2, 6
       63 | 2
       65 | 4, 4, 7
          Mean: 535.2   Median: 553   Mode: 654
   8. 1 | 2
       2 | 5, 8, 9
       3 | 5, 5, 7, 9
       5 | 2, 3, 5, 6
       6 | 8
          Mean: 40.3   Median: 37   Mode: 35
   9. 15 | 0, 5
       17 | 5, 5, 6
       22 | 7
       23 | 0, 3, 5, 5
          Mean: 199.1   Median: 201.5   Mode: 175, 235
   10. 1 | 2, 3, 3, 3, 4, 4, 5, 5, 6, 8, 9
       2 | 2, 3, 5, 5
          Mean: 17.13   Median: 15   Mode: 13
   11. 6 | 7, 7, 7, 7, 8
       7 | 1, 4, 5, 5, 8, 9, 9, 9
       8 | 1, 5
       9 | 1, 1, 2, 3, 8, 9
          Mean: 79.8   Median: 79   Mode: 67
   12. 15,000   **13.** 20.7%   **14.** Rent and Net profit
   15. Cost of merchandise: $797,080.20;   Utilities: $79,708.02;
       Rent: $119,562.03;   Maintenance: $79,708.02;   Salaries:
       $185,985.38;
       Net profit: $66,423.35
   16. $853,894.20   **17.** 30,000

Chapter 4, Unit 2

Exercise A   **1.** 25 ft. 2 in.   **2.** 26 gross 3 doz.   **3.** 28 gal.
4. 35 gal. 1 pt.   **5.** 42 gal.   **6.** 4 ft. 5 in.
7. 6 gross 10 doz.   **8.** 8 yd. 2 ft. 11 in.
9. 2 gross 8 doz. 10 units
10. 4 gross 6 doz. 11 units

Exercise B   **11.** 29 lb. 13 oz.   **12.** $88   **13.** 4 lb. 9 oz.

Chapter 4, Unit 3

Exercise A   **1.** 18 ft. 9 in.   **2.** 38 yd. 2 ft. 8 in.   **3.** 55 gal.
4. 53 lb. 10 oz.   **5.** 26 yd. 2 ft.   **6.** 2 ft.   **7.** 3 lb. 2 oz.
8. 2 gal. 2 qt.   **9.** 2 lb. 2 oz.   **10.** 2 gross 7 doz. 11 units

Exercise B   **11.** (a) 21 gross 10 doz.   (b) $157.20
12. (a) 83 lb. 4 oz.   (b) $59.70   **13.** 2 gal. 2 qt.

Chapter 4, Unit 4

Exercise A   **1.** 2.40 m   **2.** 530 mm   **3.** 0.573 L   **4.** 0.0415 L
5. 530 mg   **6.** 4.375 m   **7.** 600 mg   **8.** 80 m
9. 2,350 cg   **10.** 525 cL   **11.** 23,500 cL
12. 0.00432 kg   **13.** 50,000 mL   **14.** 400 m   **15.** 6,350 mm

Exercise B   **16.** 2.64 L   **17.** 50   **18.** 3

Chapter 4, Unit 5

Exercise A   **1.** 66 ft.   **3.** 81 mi.   **3.** 1,191 g   **4.** 71 mi.   **5.** 17 L
6. 8 lb.   **7.** 1 metric ton   **8.** 83 mm   **9.** 7 dm
10. 2 kg   **11.** 12.8 in.   **12.** 1.65 lb.

Exercise B   **13.** 1.37 in.   **14.** 4.8 L   **15.** 11.8 ft. by 21.3 ft.

Chapter 4, Unit 6

1. 31 yd. 1 ft. 6 in.   **2.** 9 gross 8 doz. 10 units   **3.** 112 gal. 1 pt.
4. 2 gross 5 doz. 10 units   **5.** 800 mg   **6.** 825 cL   **7.** 730 m
8. 1,150 mL   **9.** 1,503 g   **10.** 26 mi.   **11.** 29 mm
12. 23.26 sq. yd.

Chapter 5, Unit 1

Exercise A

	Amount of Interest	New Balance		Amount of Interest	New Balance
1.	$ 924.00	$ 9,324.00	**6.**	$ 225.00	$15,225.00
2.	605.00	11,605.00	**7.**	348.15	8,788.15
3.	609.00	6,409.00	**8.**	1,062.50	19,812.50
4.	3,400.00	13,400.00	**9.**	288.36	15,013.36
5.	2,176.53	13,941.53	**10.**	102.50	24,102.50

11. Interest Rate: 6.75%;   New Balance: $6,244.88
12. Interest Rate: 5.25%;   New Balance: $4,920.44
13. Amount: $9,750.04;   New Balance: $10,883.97
14. $3,600   **15.** $1,347.50   **16.** $993.94   **17.** $2,656.25
18. $888.96

	Amount of Interest	New Balance
19.	$ 64.14	$1,529.89
20.	101.47	2,630.27
21.	226.87	4,748.35

Exercise B   **22.** $6,648   **23.** 14.5%   **24.** $17,217.54

Chapter 5, Unit 2

Exercise A   **1.** Bills: $107:   Coins: $2.77;   Checks: $127.25;
Total Deposit: $237.02
2. Bills: $191:   Checks: $215.28;   Total Deposit: $406.28
3. Bills: $137:   Checks: $158.05;   Total Deposit: $295.05

(*Note:* For #s 4–6 follow the model check shown on page 148.)

4. Balance in register: $171.10
5. Balance in register: $357.85
6. Balance in register: $234.07

Exercise B   (*Note:* For #s 7–11 follow the model bank reconciliation statement on page 150.)

7. Adjusted balance: $245.49
8. Adjusted balance: $243.18
9. Adjusted balance: $279.90

10. Adjusted balance: $155.25
11. Adjusted balance: $558.90

Exercise C   **12.** $195.87   **13.** $57.20   **14.** $231.62   **15.** $215.80

Chapter 5, Unit 3

Exercise A   **1.** $21.25   **2.** $48.375   **3.** $63.625   **4.** $36.875   **5.** $19.75
     6. $2,690   **7.** $4,762.50   **8.** $14,450   **9.** $10,462.50
     10. $9,428.13   **11.** $18.50   **12.** $63.625   **13.** $41.875
     14. $24.625   **15.** $63.125

	Price per Share	Net Total Price
16.	$54.56	$19,096.00
17.	20.25	7,087.50
18.	7.94	2,779.00
19.	27.44	9,604.00
20.	15.25	5,337.50

Exercise B   **21.** $58.45   **22.** $19.90   **23.** $147.08   **24.** $82.94   **25.** $20.44

	Selling Price	Brokerage Fee	Total Cost
26.	$6,900.00	$125.70	$7,025.70
27.	4,375.00	116.88	4,491.88
28.	5,332.50	67.99	5,400.49

	Selling Price	Brokerage Fee	Net Proceeds
29.	$ 9,525.00	$151.73	$ 9,373.27
30.	11,400.00	190.60	11,209.40

Exercise C   **31.** (a) $10,281.25   (b) $187.50
     32. (a) $7,992.19   (b) $68.75
     33. $15,268.75
     34. (a) $5,225.00   (b) $13,775.00

Chapter 5, Unit 4

Exercise A   **1.** $935   **2.** $853.75   **3.** $1,022.50   **4.** $477.50
     5. $1,151.25   **6.** $946.25   **7.** $892.50   **8.** $951.25
     9. +$7.50   **10** –$6.25   **11.** –$21.25   **12.** +$16.25
     13. –$17.50

	Interest Rate	Maturity Year
14.	9.0%	2016
15.	7.25%	2045
16.	8.5%	2022
17.	11.75%	2003

	Price per Bond	Total Price
18.	$687.50	$3,437.50
19.	757.50	3,787.50
20.	675.00	8,100.00
21.	588.75	8,831.25

	Price per Bond	Total Price
22.	$480.00	$ 1,440.00
23.	935.00	11,220.00
24.	542.50	8,137.50
25.	940.00	12,220.00

	Price Paid	Annual Interest	Current Yield
26.	$1,022.50	$141.25	13.8%
27.	857.50	130.00	15.2%

Exercise B

	Fee per Bond	Total Fee
28.	$4.25	$21.25
29.	4.25	34.00
30.	2.75	41.25
31.	8.50	51.00
32.	6.75	94.50

	Fee per Bond	Total Fee	Total Cost
33.	$8.50	$42.50	$ 4,817.50
34.	4.25	34.00	5,624.00
35.	6.75	60.75	7,868.25
36.	2.75	41.25	7,260.00
37.	4.25	68.00	10,568.00

Exercise C **38.** (a) $10,380.00 (b) $622.80 (c) 6.9%
39. (a) $11,340.00 (b) $94.50 (c) $11,434.50

Chapter 5, Unit 5

	Amount of Interest	New Balance
1.	$1,231.88	$7,801.88
2.	2,765.34	11,105.34
3.	992.55	16,592.55
4.	161.92	4,561.92

5. Bills: $377.00; Checks: $660.55; Total Deposit: $1,037.55

6. Bills: $699.00; Checks: $956.54; Total Deposit: $1,655.54

7. Check #143 (follow model on page 148); balance is $369.59

8. Bank reconciliation statement (follow model on page 150), adjusted balance is $877.18

9. $63.125   **10.** $18.875   **11.** $31.625   **12.** $14,090.63

13. $5,443.75   **14.** $39.125   **15.** $47.25

16. Price per Share: $87.00;   Net Total Price: $41,325.00

17. $126.19   **18.** $56.03

19. Selling Price: $5,925.00;   Brokerage Fee: $149.03;   Total Cost: $6,074.03

20. Selling Price: $4,875.00;   Brokerage Fee: $99.38;   Net Proceeds: $4,775.62

21. Selling Price: $8,787.50;   Brokerage Fee: $101.09;   Net Proceeds: $8,686.41

22. $1,051.25   **23.** $957.50   **24.** –$12.50   **25.** +$6.25

26. Interest Rate: 7.5%;   Maturity Year: 2001

27. Price per Bond: $782.50;   Total Price: $6,260.00

28. Price per Bond: $613.75;   Total Price: $9,206.25

29. Price Paid: $871.25;   Annual Interest: $131.25;   Current Yield: 15.1%

30. Fee per Bond: $6.75;   Total Fee: $54.00

31. Fee per Bond: $6.75;   Total Fee: $101.25

32. Fee per Bond: $8.50;   Total Fee: $110.50;   Total Cost: $13,094.25

33. $12,153.61   **34.** $32,846.72   **35.** Corrected balance: $16.29

Chapter 6, Unit 1

Exercise A   **1.** December 19   **2.** September 1   **3.** February 20
4. February 16   **5.** September 8   **6.** April 23
7. September 14   **8.** February 18   **9.** February 28
10. August 18

	Due Date	Interest	Maturity Value
11.	October 1	$ 25.68	$815.68
12.	November 12	121.13	596.13
13.	January 29	38.00	988.00
14.	July 18	6.59	571.59
15.	June 10	18.65	783.65

Exercise B **16.** June 7; $698.63
17. September 24; $1,614.38
18. $1,200

Chapter 6, Unit 2

Exercise A

	Amount of Interest	Net Amount
1.	$ 257.00	$1,028.00
2.	50.92	2,299.08
3.	2,278.50	3,146.50
4.	377.34	2,497.66
5.	178.64	2,396.36

	Amount of Interest	Net Amount	Monthly Payment
6.	$ 365.63	$1,509.38	$104.17
7.	1,573.20	2,986.80	126.67
8.	715.00	2,035.00	114.58
9.	579.38	1,995.62	143.06
10.	2,115.00	2,385.00	93.75

11. $13.70   **12.** $24   **13.** $105.22   **14.** $64.76   **15.** $400

	Interest	Amount Due
16.	$ 69.50	$3,544.50
17.	123.15	2,586.15
18.	14.32	4,309.32
19.	64.00	2,624.00
20.	128.70	4,418.70

Exercise B **21.** $950   **22.** $2,022.47   **23.** $2,073.60   **24.** $1,390.55
25. $23.53

Chapter 6, Unit 3

Exercise A   **1.** $127   **2.** $36   **3.** $595   **4.** $208   **5.** $94
   6. $16.33   **7.** $9.55   **8.** $17.74   **9.** $19.65
   10. $21.21   **11.** 34.1%   **12.** 25.3%   **13.** 34.8%
   14. 32%   **15.** 41.7%

Exercise B   **16.** $41   **17.** $25.42   **18.** $122.60

Chapter 6, Unit 4

1. June 15   **2.** July 19   **3.** February 1

	Due Date	Interest	Maturity Value
4.	November 25	$83.69	$2,658.69
5.	July 5	28.59	1,528.59

	Interest	Total Debt
6.	$496.13	$1,846.13
7.	842.81	4,717.81
8.	912.66	5,037.66

	Interest	Net Amount	Monthly Payment
9.	$1,507.85	$4,182.15	$237.08
10.	3,400.80	3,139.20	136.25

11. $390   **12.** $28.27   **13.** $49.10

Chapter 7, Unit 1

Exercise A

	Monthly Payments	Yearly Payments
1.	$462.98	$5,555.76
2.	526.13	6,313.56
3.	752.07	9,024.84
4.	487.34	5,848.08
5.	329.61	3,955.32

6. $931.25   **7.** $997.60   **8.** $1,554.45   **9.** $109.90
10. $610.85

Exercise B   **11.** $767.10   **12.** $743.25   **13.** $42,565.22   **14.** $1,197.63
   15. $1,041.45

Chapter 7, Unit 2

Exercise A   **1.** $239.75   **2.** $179.08   **3.** $194.40   **4.** $891.41   **5.** $968

	Unexpired Days	Amount of Refund
6.	290	$147.78
7.	130	79.78
8.	185	108.97

Exercise B   **9.** $787.05   **10.** $211.68

Chapter 7, Unit 3

Exercise A   **1.** $1,375   **2.** $1,450   **3.** $1,550
4. Annual Depreciation: $2,025;   Rate of Depreciation: 21%
5. Annual Depreciation: $1,150;   Rate of Depreciation: 15%

Exercise B   **6.** $1,366.67   **7.** 18%   **8.** $1,281.25

Chapter 7, Unit 4

Exercise A   **1.** $81   **2.** $280   **3.** $84   **4.** $48   **5.** $88

Exercise B   **6.** Bodily Injury: $244;   Property Damage: $107;   Total: $351
7. $298.75   **8.** (a) $383   (b) $311

Chapter 7, Unit 5

Exercise A

	Premium per $1,000	Number of 1,000's in Face Value	Annual Premium
1.	$ 9.21	22.5	$ 207.23
2.	22.68	24.5	555.66
3.	9.33	19.5	181.94
4.	25.35	40	1,014.00
5.	10.42	29	302.18

Exercise B   **6.** $286.43   **7.** $95.37   **8.** $1,996.15

Chapter 7, Unit 6

1. Monthly Payment: $427.76;   Yearly Payment: $5,133.12
2. Monthly Payment: $527.71;   Yearly Payment: $6,332.52
3. $2,131.61   **4.** $264.38   **5.** $130.90   **6.** $188.40   **7.** $764.06

8. Unexpired Days: 252;   Amount of Refund: $162.25
9. Unexpired Days: 170;   Amount of Refund: $117.84
10. $1,006.67
11. Annual Depreciation: $1,433;   Rate of Depreciation: 14%
12. Annual Depreciation: $1,419;   Rate of Depreciation: 15%
13. $191   **14.** $81
15. Premium per $1,000: $9.21;   Number of 1,000's: 52.75;
    Annual Premium: $485.83
16. Premium per $1,000: $51.60;   Number of 1,000's: 70;
    Annual Premium: $3,612
17. (a) $607.38   (b) $33,262.80
18. (a) $3,214.69   (b) $5,024.44

Chapter 8, Unit 1

Exercise A	**1.**	$ 17.85	**2.**	$37.50	**3.**	$25.00
		35.00		29.80		35.70
		39.75		9.75		11.25
		$ 92.60		11.80		$71.95
		Tax 7.41		$88.95		Tax 2.16
		Total $100.01		Tax 6.23		Total $74.11
				Total $95.18		

Exercise B	**4.**	$55.90	**5.**	$17.50
		15.50		14.00
		7.50		11.90
		9.00		$43.40
		$87.90		Tax 1.30
		Tax 4.40		Total $44.70
		Total $92.30		

Chapter 8, Unit 2

Exercise A   **1.** $1.78   **2.** $.93   **3.** $.92   **4.** $9   **5.** $15.60

	Fractional Part	Cost of Fractional Part
6.	$\frac{3}{4}$	$4.01
7.	$\frac{7}{9}$	2.71
8.	$\frac{2}{3}$	1.43

9. $1\frac{1}{3}$ 8.33

10. $1\frac{9}{16}$ 4.45

Exercise B 11. (a) $\frac{3}{8}$ (b) $3.87

12. $3.74 13. $3.19 14. $2.35 15. $7.34

Chapter 8, Unit 3

		Discount Amount	Sale Price
Exercise A	1.	$44.07	$ 81.83
	2.	42.38	127.12
	3.	58.33	29.17

4. Remaining Percent: 70%; Sale Price: $81.03

5. Remaining Percent: $33\frac{1}{3}$%; Sale Price: $49.17

Exercise B 6. $199.31 7. $256.50 8. 25%

Chapter 8, Unit 4

1. $ 44.25
   95.70
   5.35
   ───────
   16.50
   $161.80
Tax 12.14
Total $173.94

2. $3.20 3. $4.63

4. Fractional Part: $\frac{29}{32}$; Cost: $3.49

5. Fractional Part: $1\frac{3}{4}$; Cost: $12.16

6. Discount Amount: $59.65; Sale Price: $119.30
7. Remaining Percent: 55%; Sale Price: $188.57

8. Remaining Percent: $87\frac{1}{2}$%; Sale Price: $148.71

9. $1.74 10. $747.45

Chapter 9, Unit 1

		Discount Date	Due Date
Exercise A	**1.**	April 15	May 5
	2.	May 5	June 9

		Discount Date	Due Date
	3.	February 21	April 12
	4.	September 7	October 22
	5.	January 23	February 12

		Discount Date	Due Date	Net Amount	Full Amount
	6.	9/20	10/25	$ 458.15	
	7.	7/30	9/13	380.92	
	8.	10/18	11/22		$963.85
	9.	2/18	5/9	1,217.65	
	10.	6/26	8/15	361.28	

Exercise B **11.** (a) 3/24   (b) 5/8   (c) $802.68
12. (a) 7/31   (b) 9/4   (c) $620.56
13. (a) 6/2   (b) 7/7   (c) $25.91

Chapter 9, Unit 2

		Amount of Discount	Invoice Price
Exercise A	**1.**	$ 85.93	$159.57
	2.	67.80	67.80
	3.	142.54	174.21
	4.	128.43	105.07
	5.	208.14	485.66

		First Discount	Second Discount	Third Discount	Invoice Price
	6.	$ 30.63	$ 5.69	$ 2.56	$ 48.62
	7.	67.90	3.40	1.94	62.56
	8.	42.08	9.82	4.42	83.93
	9.	74.60	34.81	6.96	132.28
	10.	127.30	35.46	20.09	180.85

11. $254.60   **12.** $324.63   **13.** $342.61   **14.** $446.78
15. $438.30   **16.** 53.3%   **17.** 50.6%   **18.** 92.5%
19. 53.0%   **20.** 54.9%

	Single Equivalent Percent	Invoice Price
21.	46%	$ 39.20
22.	48.7%	123.97
23.	44.4%	156.63
24.	46%	171.42
25.	45.9%	256.38

Exercise B
26. (a) $158.30   (b) $237.45
27. $565.40   **28.** $761.37
29. (a) A: 41.3%;   B: 40.2%   (b) A
30. $421.59

Chapter 9, Unit 3

	Markup Amount	Cost Price
Exercise A **1.**	$47.25	$87.75
2.	17.03	20.82
3.	39.38	48.12
4.	42.26	78.49
5.	73.71	90.09

6. Cost Price: $67.75;   Markup Percent: 36.2%

7. Markup Amount: $21.30;   Markup Percent: $33\frac{1}{3}\%$

8. Markup Amount: $44.95;   Markup Percent: 32.4%
9. Markup Amount: $175.60;   Markup Percent: 55%
10. Cost price: $105.70;   Markup percent: 44.8%

	Equivalent Cost Percent	Selling Price
11.	55%	$170.27
12.	60%	114.58
13.	45%	812.22
14.	70%	123.36
15.	75%	251.00

Exercise B
16. (a) $89.25   (b) 48.2%
17. (a) $292.50   (b) $117.00
18. $19.22

Chapter 9, Unit 4

		Markup Amount	Selling Price
Exercise A	**1.**	$36.56	$ 85.31
	2.	51.59	145.39
	3.	80.11	203.36
	4.	83.18	194.08
	5.	91.00	256.45

6. Selling Price: $206.88;   Markup Rate: 50.5%
7. Markup Amount: $140.15;   Markup Rate: 65.1%
8. Cost Price: $93.45;   Markup Rate: 55%
9. Selling Price: $434.78;   Markup Rate: 65%
10. Selling Price: $214.05;   Markup Rate: 70%

	Equivalent Percent	Cost Price
11.	173%	$104.48
12.	180%	240.50
13.	182%	32.01
14.	165%	63.91
15.	185%	63.51

Exercise B **16.** (a) $302.74   (b) $768.49
17. (a) $145.25   (b) 61.7%
18. (a) 175%   (b) $22.83

Chapter 9, Unit 5

1. Discount Date: March 1;   Due Date: April 20
2. Discount Date: December 15;   Due Date: March 5
3. Discount Date: 8/25;   Due Date: 9/29;   Net Amount: $857.97
4. Discount Date: 12/2;   Due Date: 1/16;   Net Amount: $1,623.66
5. Trade Discount: $282.94;   Invoice Price: $345.81
6. Trade Discount: $142.98;   Invoice Price: $214.47
7. $304.18
8. Discounts: (1) $57.18;   (2) $8.58;   (3) $3.86;   Invoice Price: $73.33
9. Discounts: (1) $222.14;   (2) $21.72;   (3) $9.99;   Invoice Price: $239.80
10. $107.17   **11.** 61.5%   **12.** 44.7%
13. Markup Amount: $64.31;   Cost Price: $119.44
14. Cost Price: $128;   Markup Percent: 45%
15. Cost Price: $249.97;   Markup Percent: 37.5%
16. Equivalent Cost Percent: 55%;   Selling Price: $450.82
17. Markup Amount: $59.25;   Selling Price: $158.00

18. Markup Amount: $188.44;   Selling Price: $478.34
19. Markup Amount: $123.25;   Markup Rate: 54%
20. $575.65
21. (a) 68.5%   (b) $103.87
22. $63.33

Chapter 10, Unit 1

Exercise A

	Regular Hours	Total Regular Amount	O/T Hours	O/T Rate	Total O/T Amount	Gross Pay
1.	40	$230.00	3	8.625	$25.88	$255.88
2.	40	180.00	1	6.75	6.75	186.75
3.	40	170.00	$7\frac{1}{4}$	6.375	46.22	216.22
4.	40	248.00	$2\frac{3}{4}$	9.30	25.58	273.58
5.	40	200.00	6	7.50	45.00	245.00
6.	37	197.95	$1\frac{1}{2}$	8.025	12.04	209.99
7.	39	304.20	$3\frac{1}{2}$	11.70	40.95	345.15
8.	37	192.40	4	7.80	31.20	223.60
9.	38	296.40	5	11.70	58.50	354.90
10.	37	175.75	$7\frac{1}{2}$	7.125	53.44	229.19

Exercise B

	Total Pieces	Total Wages
11.	589	$235.60
12.	598	199.33
13.	552	248.40
14.	580	203.00
15.	589	441.75

	Total Dozens	Total Wages
16.	$59\frac{1}{4}$	$174.79
17.	$70\frac{1}{4}$	195.30
18.	$66\frac{1}{4}$	193.45
19.	$67\frac{3}{4}$	206.64
20.	68	198.56

	Net Sales	Total Earnings
21.	$ 2,185.60	$369.13
22.	6,524.00	326.20
23.	3,478.50	353.28
24.	5,460.00	313.40
25.	12,760.00	382.80

Exercise C   **26.** $333.28   **27.** $415.01   **28.** $313.03   **29.** $100,100
30. $171.45   **31.** $213.75   **32.** $96.68   **33.** 22.7%

Chapter 10, Unit 2

Exercise A   **1.** $36.38   **2.** $34.69   **3.** $35.47   **4.** $29.51   **5.** $44.05
6. none   **7.** none   **8.** 47   **9.** 40   **10.** 33

Exercise B

	Income Tax	FICA Tax	Take-Home Pay
11.	$72.00	$58.71	$636.79
12.	68.00	60.72	665.03
13.	127.00	60.09	598.41
14.	72.00	66.82	734.63
15.	63.00	52.08	565.67

Exercise C   **16.** $518.06
17. $52,469.04
18. $738.68

Chapter 10, Unit 3

	Regular Pay	O/T Pay	Gross Pay
1.	$310.00	$ 69.75	$379.75
2.	274.00	123.30	397.30
3.	306.00	88.93	394.93

4. Total Pieces: 459;   Total Wages: $275.40
5. Total Pieces: 1,146;   Total Wages: $401.10
6. Total Dozens: 125;   Total Wages: $316.25
7. $35.48   **8.** $43.51   **9.** 51   **10.** 44
11. Income Tax: $64.00;   FICA: $58.40;   Take-Home Pay: $640.97
12. Income Tax: $75.00;   FICA: $68.65;   Take-Home Pay: $753.77
13. $95   **14.** $383.48

Chapter 11, Unit 1

Exercise A   **1.** Total Assets: $132,524.00;   Total Liabilities: $61,570.00
   Equity (net worth): $70,954.00
   2. Total Assets: $119,669.00;   Total Liabilities: $68,838.00
   Equity (net worth): $50,831.00

Word Problems
Exercise B   **1.** Assets: $585,017.00;   Liabilities: $314,417.00;
   Equity (net worth): $270,600.00

   2. Assets: $77,156.00;   Liabilities: $59,112.00;
   Equity (net worth): $18,044.00
   3. Assets: $677,280.00;   Liabilities: $100,200.00;
   Equity (net worth): $577,080.00
   4. Assets: $121,366.00;   Liabilities: $67,340.00;
   Equity (net worth): $54,026.00
   5. Assets: $490,641.00;   Liabilities: $293,030.00
   Equity (net worth): $197,611.00

Chapter 11, Unit 2

Exercise A   **1.** Total Revenues: $16,826.45;   Total Liabilities: $19,032.00;
Net Profit: $(−) 2,205.55
2. Total Revenues: $15,712.75;   Total Liabilities: $7,231.00;
Net Profit: $8,481.75
3. Total Revenues: $15,875.00;   Total Liabilities: $13,282.00;
Net Profit: $2,593.00

Word Problems

Exercise B   **1.** Revenues: $13,576.00;   Liabilities: $12,513.00;
Net Profit: $1,063.00
2. Revenues: $7,541.00;   Liabilities: $7,781.00;
Net Profit: (−) $240.00 net loss
3. Revenues: $22,916.00;   Liabilities: $19,207.00;
Net Profit: $3,709.00
4. Revenues: $10, 576.00;   Liabilities: $9,163.00;
Net Profit: $1,413.00
5. Revenues: $23,218.00;   Liabilities: $19,268.00;
Net Profit: $3,950.00

Chapter 11, Unit 3

1. $5,010.00	**2.** $3,390.00	**3.** $3,382.84	**4.** $3,577.00
5. $2,424.00	**6.** $2,396.59	**7.** $42,184.71	**8.** $58,407.05
9. 2	**10.** 3		

Chapter 11, Unit 4

1. 5,375   **2.** $5,312.50   **3.** $49.00   **4.** $8,285.71

5. Year	Fraction	Book Value
1	$\frac{9}{45} \times 18,000 = 3,600$	$14,400.00
2	$\frac{8}{45} \times 18,000 = 3,200$	$11,200.00

Year	Fraction	Book Value
3	$\dfrac{7}{45} \times 18{,}000 = 2{,}800$	$8,400.00
4	$\dfrac{6}{45} \times 18{,}000 = 2{,}400$	$6,000.00
5	$\dfrac{5}{45} \times 18{,}000 = 2000$	$4,000.00
6	$\dfrac{4}{45} \times 18{,}000 = 1{,}600$	$2,400.00
7	$\dfrac{3}{45} \times 18{,}000 = 1{,}200$	$1,200.00
8	$\dfrac{2}{45} \times 18{,}000 = 800$	$400.00
9	$\dfrac{1}{4} \times 18{,}000 = 400$	$0.00

6.

Year	Fraction	Book Value
1	$\dfrac{12}{78} \times 17{,}500 = 2{,}692.31$	$14,807.29
2	$\dfrac{11}{78} \times 17{,}500 = 2{,}467.95$	$12,339.34
3	$\dfrac{10}{78} \times 17{,}500 = 2{,}43.57$	$10,095.77
4	$\dfrac{9}{78} \times 17{,}500 = 2{,}019.23$	$8,076.54
5	$\dfrac{8}{78} \times 17{,}500 = 1{,}794.87$	$6,281.67
6	$\dfrac{7}{78} \times 17{,}500 = 1{,}570.51$	$4,711.16
7	$\dfrac{6}{78} \times 17{,}500 = 1{,}346.15$	$3,365.01
8	$\dfrac{5}{78} \times 17{,}500 = 1{,}121.79$	$2,243.22

Year	Fraction	Book Value
9	$\frac{4}{78} \times 17{,}500 = 879.44$	$1,345.78
10	$\frac{3}{78} \times 17{,}500 = 673.08$	$672.70
11	$\frac{2}{78} \times 17{,}500 = 448.72$	$223.98
12	$\frac{1}{78} \times 17{,}500 = 224.35$	$0.00

7.

Year	Fraction	Book Value
1	$\frac{9}{45} \times 9{,}500 = 1{,}900$	$7,600.00
2	$\frac{8}{45} \times 9{,}500 = 1{,}688.89$	$5,911.11
3	$\frac{7}{45} \times 9{,}500 = 1{,}477.78$	$4,433.33
4	$\frac{6}{45} \times 9{,}500 = 1{,}266.67$	$3,166.66
5	$\frac{5}{45} \times 9{,}500 = 1{,}055.56$	$2,111.10
6	$\frac{4}{45} \times 9{,}500 = 844.44$	$1,266.66
7	$\frac{3}{45} \times 9{,}500 = 633.33$	$633.33
8	$\frac{2}{45} \times 9{,}500 = 422.22$	$211.11
9	$\frac{1}{45} \times 9{,}500 = 211.11$	$0.00

8.

Year	Depreciation	Book Value
1	$0.2 \times 10{,}500.00 = \$2100.00$	$10{,}500 - 2{,}100 = \$8{,}400.00$
2	$0.2 \times 8{,}400.00 = \$1680.00$	$8{,}400 - 1{,}680 = \$6{,}720.00$
3	$0.2 \times 6{,}720.00 = \$1344.00$	$6{,}720 - 1{,}344 = \$5{,}376.00$
4	$0.2 \times 5{,}376.00 = \$1075.20$	$5{,}376 - 1{,}075.20 = \$4{,}300.80$
5	$0.2 \times 4{,}300.80 = \$860.16$	$4{,}300.80 - 860.16 = \$3{,}440.64$
6	$0.2 \times 3{,}440.64 = \$688.13$	$3{,}440.64 - 688.13 = \$2{,}752.51$

Year	Depreciation	Book Value
7	$0.2 \times 2{,}752.51 = \$550.50$	$2{,}752.51 - 550.50 = \$2{,}202.01$
8	$0.2 \times 2{,}202.01 = \$440.40$	$2{,}202.01 - 440.40 = \$1{,}761.61$
9	$0.2 \times 1{,}761.61 = \$352.32$	$1{,}761.61 - 352.32 = \$1{,}409.29$
10	$0.2 \times 1{,}409.29 = \$281.86$	$1{,}409.29 - 281.86 = \$1{,}127.43$

9.

Year	Depreciation	Book Value
1	$0.25 \times 105{,}000.00 = \$26{,}250.00$	$105{,}000 - 26{,}250 = \$78{,}750.00$
2	$0.25 \times 78{,}750.00 = \$19{,}687.50$	$78{,}750 - 19{,}687.50 = \$59{,}062.50$
3	$0.25 \times 59{,}062.50 = \$14{,}765.63$	$59{,}062.50 - 14{,}765.63 = \$44{,}296.87$
4	$0.25 \times 44{,}296.87 = \$11{,}074.22$	$44{,}296.87 - 11{,}074.22 = \$33{,}222.65$
5	$0.25 \times 33{,}222.65 = \$8{,}305.66$	$33{,}222.65 - 8{,}305.66 = \$24{,}916.99$
6	$0.25 \times 24{,}916.99 = \$6{,}229.25$	$24{,}916.99 - 6{,}229.25 = \$18{,}687.74$
7	$0.25 \times 18{,}687.74 = \$4{,}671.94$	$18{,}687.74 - 4{,}671.94 = \$14{,}015.80$
8	$0.25 \times 14{,}015.80 = \$3{,}503.95$	$14{,}015.80 - 3{,}503.95 = \$10{,}511.85$

10.

Year	Depreciation	Book Value
1	$0.22 \times 18{,}500.00 = \$4{,}070.00$	$18{,}500 - 4{,}070 = \$14{,}430.00$
2	$0.22 \times 14{,}430.00 = \$3{,}174.60$	$14{,}430 - 3{,}174.60 = \$11{,}255.40$
3	$0.22 \times 11{,}255.40 = \$2{,}476.19$	$11{,}255.40 - 2{,}476.19 = \$8{,}779.21$
4	$0.22 \times 8{,}779.21 = \$1{,}931.43$	$8{,}779.21 - 1{,}931.43 = \$6{,}847.78$
5	$0.22 \times 6{,}847.78 = \$1{,}506.51$	$6{,}847.78 - 1{,}506.51 = \$5{,}341.27$
6	$0.22 \times 5{,}341.27 = \$1{,}175.08$	$5{,}341.27 - 1{,}175.08 = \$4{,}166.19$
7	$0.22 \times 4{,}166.19 = \$916.56$	$4{,}166.19 - 916.56 = \$3{,}249.63$
8	$0.22 \times 3{,}249.63 = \$714.92$	$3{,}249.63 - 714.92 = \$2{,}534.71$
9	$0.22 \times 2{,}534.71 = \$557.64$	$2{,}534.71 - 557.64 = \$1{,}977.07$

Chapter 11, Unit 5

1. Assets: $114,263;   Liabilities: $ 87,300;   Equity (net worth): $26,963

2. Assets: $226,013;   Liabilities: $151,341;   Equity (net worth): $74,672

3. Revenues:

Rental Income:	$22,450
Total Revenues:	$22,450

Expenses:

Mortgage:	$6,550
Insurance:	$1,030
Utilities:	$1,575
Building Repairs:	$1,235
Total Expenses:	$10,390

Net profit = $22,450 – $10,390 = $12,060

4. Revenues:

	Gross Sales:	$32,655
	Total Revenues:	$12,655

Expenses:

	Rent:	$1,200
	Insurance:	$ 135
	Utilities:	$ 432
	Incidentals:	$ 752
	Total Expenses:	$2,519

Net profit = $12,655 − $2,519 = $10,136

5. $1,215.00   **6.** $807.00   **7.** $799.67   **8.** $9,288.00
9. $5,340.00   **10.** $5277.10   **11.** $54,870.18
12. $48,361.90   **13.** 4   **14.** $3,650.00   **15.** $5.777.78
16. $6,833.33

17.

Year	Fraction	Book Value
1	$\frac{7}{28} \times 29,460 = 7,365$	$22,095.00
2	$\frac{6}{28} \times 29,460 = 6,312.86$	$15,782.14
3	$\frac{5}{28} \times 29,460 = 5,260.71$	$10,521.43
4	$\frac{4}{28} \times 29,460 = 4,208.57$	$6,312.86
5	$\frac{3}{28} \times 29,460 = 3,156.43$	$3,156.43
6	$\frac{2}{28} \times 29,460 = 2,104.29$	$1,052.14
7	$\frac{1}{28} \times 29,460 = 1,052.14$	$ 0.00

18.

Year	Fraction	Book Value
1	$\frac{6}{21} \times 67,500 = 19,285.71$	$48,214.29
2	$\frac{5}{21} \times 67,500 = 16,071.43$	$32,142.86
3	$\frac{4}{21} \times 67,500 = 12,857.14$	$19,285.72
4	$\frac{3}{21} \times 67,500 = 9,642.86$	$9,642.86
5	$\frac{2}{21} \times 67,500 = 6,428.57$	$3,214.29
6	$\frac{1}{21} \times 67,500 = 3,214.29$	$ 0.00

19.

Year	Fraction	Book Value
1	$\frac{9}{45} \times 19,065 = 3,813$	$15,252.00
2	$\frac{8}{45} \times 19,065 = 3,389.33$	$11,862.67
3	$\frac{7}{45} \times 19,065 = 2,965.67$	$8,897.00
4	$\frac{6}{45} \times 19,065 = 2,542$	$6,355.00
5	$\frac{5}{45} \times 19,065 = 2,118.33$	$4,236.67
6	$\frac{4}{45} \times 19,065 = 1,694.67$	$2,542.00
7	$\frac{3}{45} \times 19,065 = 1,271$	$1,271.00
8	$\frac{2}{45} \times 19,065 = 847.33$	$423.67
9	$\frac{1}{45} \times 19,065 = 423.67$	$0.00

20.

Year	Depreciation	Book Value
1	$0.13 \times 26{,}700.00 = \$3{,}471.00$	$26{,}700 - 3{,}471 = \$23{,}229.00$
2	$0.13 \times 23{,}229.00 = \$3{,}019.77$	$23{,}229 - 3{,}019.771 = \$20{,}209.23$
3	$0.13 \times 20{,}209.23 = \$2{,}627.20$	$20{,}209.23 - 2{,}627.20 = \$17{,}582.03$
4	$0.13 \times 17{,}582.03 = \$2{,}285.66$	$17{,}582.03 - 2{,}285.66 = \$15{,}296.37$
5	$0.13 \times 15{,}296.37 = \$1{,}988.53$	$15{,}296.37 - 1{,}988.53 = \$13{,}307.84$
6	$0.13 \times 13{,}307.84 - \$1{,}730.02$	$13{,}307.84 \quad 1{,}730.02 = \$11{,}577.82$
7	$0.13 \times 11{,}577.82 = \$1{,}505.12$	$11{,}577.82 - 550.50 = \$10{,}072.70$
8	$0.13 \times 10{,}072.70 = \$1{,}309.45$	$10{,}072.70 - 1{,}309.45 = \$8{,}763.25$
9	$0.13 \times 8{,}763.25 = \$1{,}139.22$	$8{,}763.25 - 1{,}139.22 = \$7{,}624.03$
10	$0.13 \times 7{,}624.03 = \$991.12$	$7{,}624.03 - 991.12 = \$6{,}632.91$
11	$0.13 \times 6{,}632.91 = \$862.28$	$6{,}632.91 - 862.28 = \$5{,}770.63$
12	$0.13 \times 5{,}770.63 = \$750.18$	$5{,}770.63 - 750.18 = \$5{,}020.45$
13	$0.13 \times 5{,}020.45 = \$652.66$	$5{,}020.45 - 652.66 = \$4{,}367.79$
14	$0.13 \times 4{,}367.79 = \$567.81$	$4{,}367.79 - 567.81 = \$3{,}799.98$
15	$0.13 \times 3{,}799.98 = \$494.00$	$3{,}799.98 - 494.00 = \$3{,}305.98$

21.

Year	Depreciation	Book Value
1	$0.17 \times 31{,}250.00 = \$5{,}312.50$	$31{,}250 - 5{,}312.50 = \$25{,}937.50$
2	$0.17 \times 25{,}937.50 = \$4{,}409.38$	$25{,}937.50 - 4{,}409.38 = \$21{,}528.12$
3	$0.17 \times 21{,}528.12 = \$3{,}659.78$	$21{,}528.12 - 3{,}659.78 = \$17{,}868.34$
4	$0.17 \times 17{,}868.34 = \$3{,}037.62$	$17{,}868.34 - 3{,}037.62 = \$14{,}830.72$
5	$0.17 \times 14{,}830.72 - \$2{,}521.22$	$14{,}830.72 - 2{,}521.22 = \$12{,}309.50$
6	$0.17 \times 12{,}309.50 = \$2{,}092.62$	$12{,}309.50 - 2{,}092.62 = \$10{,}216.88$
7	$0.17 \times 10{,}216.88 = \$1{,}736.87$	$10{,}216.88 - 1{,}736.87 - \$8{,}480.01$
8	$0.17 \times 8{,}480.01 = \$1{,}441.60$	$8{,}480.01 - 1441.60 = \$7{,}038.41$
9	$0.17 \times 7{,}038.41 = \$1{,}196.53$	$7{,}038.41 - 1{,}196.53 = \$5{,}841.88$
10	$0.17 \times 5{,}841.88 = \$993.12$	$5{,}841.88 - 993.12 = \$4{,}848.76$
11	$0.17 \times 824.76 = \$824.29$	$4{,}848.76 - 824.29 = \$4{,}024.47$
12	$0.17 \times 4{,}024.47 = \$684.16$	$4{,}024.47 - 684.16 = \$3{,}340.31$